THE TECHNOLOGY OF POWDER COATINGS

by

S. T. HARRIS
B.Sc.

Edited by

E. C. ROBERSON
Ph.D., B.Sc., A.R.I.C.

PORTCULLIS PRESS
LONDON

1976

First published in 1976 by Portcullis Press Ltd.
Queensway House, 2 Queensway, Redhill, Surrey, RH1 1QS

ISBN 90199493 6

A Portcullis Press Publication

Set and printed in Gt Britain by Page Bros (Norwich) Ltd, Norwich

Preface

At their present stage of development, powder coatings must seem to the newcomer to be a confusion of plastics and paint technologies. I have, therefore, endeavoured to present a factual account of the techniques currently used in the industry and, where possible, to highlight the aspects which need to be developed.

The technology of powder coatings will alter considerably during the next decade but it will surely become established as a technology in its own right. In compiling this account I am deeply indebted to the suppliers of raw materials, manufacturing machinery and application and testing equipment who have assisted me with descriptions of their products and photographs to illustrate the text.

A list of these suppliers is given in the final section of the book to enable the reader to obtain up-to-date information on the products described in the text.

The manner in which this book is presented owes much to the careful editing of Dr E. C. Roberson and I am extremely grateful to him for easing the burden on a first time author.

The Old Rectory
Nether Whitacre
Near Coleshill
West Midlands

S. T. Harris
August 1976

Foreword

Powder coatings designed for use in industry, have now reached an acceptance level at which they can be regarded as a viable form of industrial finishing.

There comes a time in any new technology such as this when a considerable amount of information is available, acquired both by ad hoc methods and by organised research, but when no one has either enough knowledge or the dedication to put this information into a collected form and so allow both the practitioner and the newcomer to the industry to understand the status which has been reached and the possibilities for the future.

The powder coating industry in its present form has grown in a remarkable way in the last ten years. Appreciable quantities of thermoplasts were used even ten years ago, when the thinner self-adherent coatings, offered by the thermoset powders, were just coming into use. In those early days they were used mainly to give special protective properties, improved resistance to wear and damage or because of their good electrical properties, but their appearance in general was rather poor. Tremendous efforts by resin manufacturers, powder formulators and manufacturers together with enormous improvements in electrostatic spraying equipment, have now made these thermoset powder-coatings comparable in appearance with the best of the liquid coatings while still offering considerable advantages in performance.

Other factors, such as the enormous increase in recent years in the cost of hydrocarbon and organic solvents generally, have made powder coatings compare favourably in cost with liquid coatings. Environmental, safety and health considerations also make them increasingly attractive to manufacturers who need to apply coatings for protective and aesthetic reasons.

The author, whom I have known personally for many years, has a life long experience in the surface coatings field. This has included in recent years a period of dedication to powder coatings. Not only a scientist and a formulator of powder coatings, he has also had a wealth of experience of manufacture, setting up two plants for this purpose, the second being his own company.

It is therefore with considerable pleasure that I commend this book to whoever reads it. It is a fascinating study of the development of a new technology, done with considerable care and devotion and presented in a clear and organised manner. The fact that powder technology has still a long way to travel in no way detracts from this book, which should remain a standard book of reference for years to come.

Drynamels Limited

F. C. J. Ruzicka
Ph.D., D.I.C., A.R.C.S., F.R.I.C.
Managing Director

Contents

List of Tables

1

Introduction

SUMMARY

In these days of escalating raw material prices, and a worldwide shortage of the essential components required by the Paint Industry, it is not surprising that the new techniques, which improve the overall efficiency of painting methods, should now be considered and introduced as replacements for many conventional industrial finishing processes.

Already the new techniques that have been introduced on an industrial scale include electrophoretic painting, the use of water borne primers and top coats, and non-aqueous dispersions (NAD). These methods have one common purpose, namely, to reduce the dependence of organic coatings upon high proportions of costly organic diluents and solvents which have to be wasted in the drying process. The need to improve coating efficiency and to reduce the volatile content of paints has also been prompted by the desire to reduce environmental pollution caused by the disposal of these organic wastes.

It is little more than ten years since spray applied powder coatings were introduced, and it is this technical development which is fully described in the following pages although reference will be made, from time to time, to comparative assessments of powder coatings and conventional paints.

HISTORICAL BACKGROUND

Thermosetting powder coatings were introduced in the United States during the late 1950's. They were relatively simple products, dry blends of epoxy resin, pigments and hardener, dispersed by ball milling techniques. The fluidized bed method for applying thermoplastic powders was already well established at the time and the new thermosetting products were initially applied in the same manner. But due to the heterogeneous nature of the product, the individual components were easily classified on a size and weight basis in the fluidized bed and the results were inconsistent. At this stage of development thermoset powders were not considered as potential decorative coatings, but rather as chemically resistant or electrically insulating coatings with film thicknesses, in excess of two hundred microns.

In the early 1960's the Shell Research laboratories at Egham began the first serious investigation into better methods of compounding these products, and they described two methods for formulating thermosetting epoxy based powder coatings. In the first method, liquid epoxy resins and liquid hardeners were reacted to produce a partially cured B stage in which the liquid components were polymerised to a low melting point solid which could then be comminuted to a fine powder. The fully cured C stage was achieved by stoving the B stage powders at high temperature. It was, however, extremely difficult to ensure good batch to batch reproducibility, for the properties of the B stage were sensitive to small temperature variations and at low temperatures the reaction would take an excessively long time to proceed to the B stage. The most serious drawback however was the exotherm which could occur when the compounded materials were cast in too thick a layer, resulting in rapid cross linking to form the C stage. These products were also characterized by short shelf life. While B stage powders are no longer produced commercially for the reasons stated, it must be said that thickness for thickness they gave cured films of exceptional chemical resistance which cannot be matched by current powder coating systems.

The second method investigated by the Shell laboratories was termed the fusion process where solid epoxy resins, pigments and curing agent (dicyandiamide) were compounded in a heated Z-blade mixer. This compounding process was an immense step forward in the development of thermosetting powder coatings but it still had certain major faults. It was essentially a batch production process requiring excessive cleaning operations between batches, and the performance of the finished product was limited by the low reactivity of the systems manufactured by Z-blading. The use of heat activated curing agents, such as boron trifluoride complexes was partly successful in producing systems curing at lower temperatures but the problems of cleaning between batches is still a major restriction to the use of the Z-blade mixer.

Continuing research at Shell resulted in 1964 in the introduction of extrusion methods and it is this process which is currently favoured for the large scale continuous compounding of powder coatings. The high cost of the equipment deterred many powder manufacturers and the small paint producer, contemplating his own powder production facility, often proceeded no further than obtaining the quotation for the extruder.

The application of powder coatings by spray methods to non-preheated articles became a reality in 1962 with the introduction by S.A.M.E.S. of electrostatic powder spray guns. Early and cumbersome hand operated equipment was progressively improved to give light-weight guns which could be easily automated, and many novel methods for the electrostatic application of powder coatings are to be found in recent patent literature. Rapid development in application equipment over the past five years, and the gradual acceptance by industry that powder coating must be considered as a separate process from conventional painting methods, have been the main factors in the rapid growth in powder production over this period.

TYPES OF POWDER COATING

There are two main classes of powder coating; the thermoplastic and the thermosetting types. The resins used in both types have been employed in solution form for many years by the Paint Industry, and a number of the more successful coatings have resulted from the combination of compatible resins of both series.

Thermoplastic resins are generally characterized as materials of comparatively high molecular weight which exhibit excellent chemical resistance qualities, toughness and flexibility. Thermosetting resins are initially of lower molecular weight and a high molecular weight complex structure is built up during the curing cycle. As low molecular weight materials they show better flow properties, good surface wetting characteristics and hence, excellent adhesion to the substratum.

Traditionally, thermoplastic resins were well established in the field of powder coatings before thermoset products were developed. In the fluidized bed method of application, thermoplastic powders would melt in contact with the heated article to be coated, and flow out to give a continuous film upon cooling. Spoiled articles could be repaired by reheating and reflowing the film. However, application by this technique resulted in film thicknesses greater than 250 microns and the high raw material costs of some resins made this process of interest only for specialized applications. In the electrical industry a considerable tonnage of thermoplastic powder coating is used for electrical insulation purposes, a field which continues to grow at a healthy rate.

High molecular weight resins exhibiting toughness and flexibility are difficult to reduce in particle size without resorting to the costly techniques of cryogenic fine grinding. It is this factor which makes it difficult to process thermoplastic resins, such as nylon, to the particle size suitable for thin film electrostatic spray application.

Thermoplastic resins suitable for use in powder coatings include polyvinyl chloride, polyvinyl butyral, nylons, polyesters, acrylics and certain high molecular weight epoxies.

The list of thermosetting resins suitable for use in powder coatings includes heat convertible forms of many in the thermoplastic class but one resin alone is predominant, namely the epoxy resin. More than 90% of all thermosetting powder coatings are based upon the epoxy resin, and it is predicted that this domination will continue for several years.

In the period of development since 1970, polyester resins have been introduced as alternatives to epoxy resins, and acrylic resins are also available.

There are several reasons why epoxy resins are widely used in powder coatings. In the development era they already existed as coating resins possessing relatively low melting points with glass transition temperatures high enough to prevent the fine particles agglomerating at room temperatures. Their most important attributes are excellent adhesion and very good chemical resistance. Epoxy resins are capable of reacting with a wide range of organic compounds so that particular film properties may be selected by the correct choice of curing agent.

Epoxy resins are open to criticism with respect to heat and light stability and they also exhibit a tendency to chalking on exposure outdoors. Polyester resins and acrylics show distinct advantages over the epoxies in their improved heat and light resistance qualities and some of these resins are markedly better in respect of exterior weathering properties.

It is possible to cross-link hydroxyl-containing resins with an isocyanate to produce a polyurethane coating. This method, however involves the use of a blocked isocyanate in which the volatile component is driven off during the stoving operation. This volatile content defeats the main objective of powder coatings which should be the attainment of a 100% non-volatile system.

Thermosetting vinyl systems have also been developed and phenolic resins may be used in conjunction with epoxy resins. Amino resins may be included as co-reactants with polyesters and acrylics.

Almost daily, new resins are developed for use in powder coatings and the formulating techniques of the paint technologist will, no doubt, be readily applied to powder coatings.

MARKET FOR POWDER COATINGS

During the period 1969–1976 there have been many conferences dealing with powder coatings and inumerable statistics have been presented to illustrate the growth rate in this new market. While these figures have at times appeared to be contradictory there is no doubt that considerable growth has occurred during this period. In the early 1970's an annual growth rate of 25–30% was reported in many countries. This rate has now increased to between 50 and 100% and it is expected to maintain this for the next 5–6 years.

The assessment of the market potential must begin with an appreciation of the factors which are responsible for this high growth rate:

(a) In the United States and Japan, industry must comply with the terms of stringent anti-pollution legislation, while in all countries there is a growing awareness of the need to protect the environment. Powder coatings cause virtually no atmospheric pollution, nor do they produce effluent. In fact they are generally free from any toxic hazard.

(b) The current shortages of all chemicals derived from petroleum, and the escalating price of paint solvents, make it all the more desirable to consider formulating paints which do not require solvents either during manufacture or application.

(c) There are distinct advantages to be gained, both financially and on the production line by the use of a one coat dry coating to replace a two coat wet process.

(d) The fluidized bed is a simple process to operate, and the ease of application of powder coatings by electrostatic spray to complex articles removes the need to employ highly skilled applicators.
(e) Application processes for powder coatings are easy to automate and the costs of equipment are reasonable.
(f) New developments with polyester, polyurethane and acrylic resins are increasing the scope of powder coatings.
(g) The fire hazard associated with wet paint methods is considerably reduced.
(h) In practice, the efficiency of usage of powder coatings in automated plants is 95–98%.

Manufacturers who have introduced powder coatings have invariably recovered the capital cost of the plant within three years by the considerable savings effected in the following areas:

1. Primer coats are eliminated in most cases.
2. Overspray losses are greatly reduced.
3. Labour costs may be cut by as much as 50%.
4. Cleaning of the wet paint booth is eliminated.
5. Powder coatings are supplied ready for use and do not require thinning or mixing before application.
6. Factory or paint shop heating costs are reduced for it is not necessary to exhaust large quantities of warm air from the plant.
7. Reject rates are low.
8. The plant occupies less space since there is no flash-off zone and the oven may be reduced in size.
9. Costs of paint storage may be reduced.
10. Dust proof rather than flame proof equipment is specified.
11. Insurance rates should be lower, although this does not always apply when a new process is assessed by insurers.

The Size of the Market

A world wide consumption of powder coatings of all types has been estimated by De Bell and Richardson.[1]

	1971	*1976*	*1980*
Total usage (metric tonnes)	44,500	105,000	208,000

An assessment of the position in Japan was given by Hikasa.[2]

	1971	*1972*	*1973*	*1974*	*1975*
Usage (metric tonnes)	1,200	2,400	3,700	6,100	9,000

Even this significant growth will only account for 0·6% of the total Japanese paint sales by 1975.

Properties

Good adhesion, scratch resistance in excess of 4,000 gm.

High gloss and good flow.

Mainly white or pastel shades.

Light and heat stability properties may only be provided by acrylic or polyester powders.

Stain resistance qualities are met by epoxy powders.

Good alkali and humidity resistance in excess of 2000 hrs. Epoxy powders are most suitable.

Substratum may be mild steel, zinc coated steel or phenolic plastic.

Outlets in the food industry may require the coatings to have F.D.A. status.

AUTOMOTIVE

Major outlets are: bus and lorry chassis, wheels, primer surfacer, interior trim, steering columns, bumpers, rocker box covers, brake cylinders, seating frames.

Properties

The requirements will vary with the end use but high corrosion resistance is a prime feature.

Resistance to A.S.T.M. salt spray in excess of 2,500 hr.

100% humidity resistance (without impairment of adhesion and flexibility characteristics) after 2,500 hr.

Scratch resistance in excess of 4000 gm. and resistance to gravelometer and chipping tests.

Gloss requirements may be as low as 15% for primers with high gloss for one coat finishes.

Epoxy based powder coatings are suitable for all the above outlets.

ELECTRICAL

Major outlets are: switch gear cabinets, bus bar insulation, coil windings, resistors and printed circuits.

Properties

Good insulative properties are required at film thicknesses of around 15 thousandths of an inch (380 microns)

Fluidized bed application (or a special modification), is the most common coating method used to obtain the high film thicknesses.

Epoxy coatings are suitable for most electrical applications.

EXTERIOR QUALITY

Major outlets are: metal window frames, garden furniture, cycles, street lighting poles, construction metal work, siding enamels.

Properties

Good exterior weathering properties are required with, in most cases, freedom from chalking. Epoxy powder coatings show a marked tendency to chalk, so that the polyester and acrylic powders will continue to dominate this market.

Nylon coatings are widely used for many exterior applications.

GENERAL ENGINEERING

Major outlets are: fire extinguishers, display equipment, shelving, partitioning, tools, castings.

This field is currently the largest sector for powder coatings where the main requirements are toughness and adhesion.

A wide range of colours, at different gloss levels make it difficult for powder manufacturers to adhere to a standard range of products.

Requirements also include hammer and texture finishes.

TRADE COATERS

It is difficult to assess the end use applications of the miscellany of products coated by this industry.

Most applications are by hand-operated electrostatic spray and efficient collection of oversprayed powder is not always possible. A major problem, therefore, is the possibility of cross contamination between different types of powder and, for this reason, most trade coaters prefer to use the standard range of epoxy based powder coatings.

Market Opportunities

There are several market opportunities for powder coatings which are already well established or currently being developed.

POWDER COATING OF GLASS BOTTLES

A clear coating may be applied to glass bottles to make them shatter proof. In the U.S.A. this is attracting attention because of proposed federal regulations to increase the safety of handling pressurized beer and beverages in bottles. An economic incentive is the possibility of reducing the weight of glass.

Assuming a coating thickness of 0·004 to 0·005 inches (0·1 to 0·15 mm), the potential market in the U.S.A. for the soft drinks industry alone is predicted at 90,000 tonnes.(5)

Both thermoset and thermoplastic resins have been considered including epoxies, vinyls, nylon, polyethylene, polypropylene, proprionates and butyrates. The coating thickness could be achieved by electrostatic spray or fluidized bed.

An allied market exists for coating non pressurized bottles for cosmetics with opaque and self coloured transparent films.

PIPE COATINGS

A market that is already well established in the U.S.A. and is considered to be a major growth area in Europe.

In 1972 the U.S. market for epoxy based pipe coatings was 2,500 tonnes while the total usage in Europe was less than 100 tonnes. By the end of 1974 it is predicted that European demand will show a tenfold increase.[6]

The requirements for a good pipe coating are;

The coating must be non conductive and impervious to the passage of water through the film.

Good impact resistance and tensile strength to resist the weight of earth on the pipe. Good flexibility and adhesion, even at low temperatures, so that the pipe may be bent.

Resistance to cathodic disbondment.

The pipes are often coated straight from the production line and the coating is cured by residual heat. Small bore pipes require a powder with a fast curing cycle.

STEEL CONTAINERS

An interesting example of on site coating of large containers has been given by Hideo Nagasaka,[7] in which large closed vessels were coated with epoxy powders applied by flame spraying methods. Advantages claimed were, freedom from solvent taint in food containers, ease of application with no need to post cure, and high film thickness in one coat.

CAN COATINGS

The film thickness of coatings used in this industry lie between 0·004 to 0·008 mm for internal coatings and 0·015 to 0·025 mm for exterior coatings. The major problems of introducing powder coatings for the above outlets will be; speed of application and uniformity of film, inadequate film flow at low coating weights, difficulties of producing bit-free powders due to dust collection during manufacture.

However, powder coatings are currently being used in the U.S.A. as side stripe coatings for protecting the lapped seam and this requirement could grow with the wider acceptance of tin free steel.

A possible outlet for powder coatings is the field of "hard to hold" products, where the container has to withstand the corrosive action of its contents—e.g. emulsion paints, chemical cleaners.

DRUM COATINGS

The steel drum may be coated easily with epoxy based powder coatings which are resistant to attack by the range of products generally packed in drums. A major problem could occur between the application bay and the oven if the coated drums were subjected to any vibration in transit, this could dislodge the electrostatically applied powder coatings. The main advantage, however, would be the replacement of the present two coat wet-on-wet operation by a one coat process thereby increasing the speed of production. Further advantages are the excellent coverage of powder along the seam of the drum and the marked improvement in adhesion and flexibility of the cured film.

CONCLUSION

There is little doubt that powder coatings, already firmly established in many of the traditional paint markets, will continue to grow rapidly during the next decade. The extent of their penetration into the industrial paint field will depend largely upon their acceptance by the automotive and domestic appliance industries.

REFERENCES

1. Childs and Holappa (De Bell and Richardson, Connecticut, U.S.A.). *Pigment and Resin Technology* (October, 1972).
2. Akira Hikasa (Nippon Paint Company). "A New Development in Thermosetting Acrylic Powder Coatings". 4th International Powder Symposium, London (1974).
3. Dr. S. Kut. "Developments in Thermosetting Powder Coatings", 4th International Powder Symposium, London (1974).
4. Bruniere (ATO Plastique, Paris). International Symposium, Brussels (1973).
5. *Modern Plastics International* (May, 1973).
6. C. Buhler (Dow Chemicals, Europe SA, Switzerland).
7. Professor Hideo Nagasaki. "Flame Spraying of Powder Coatings", 4th International Powder Symposium, London (1974).

2

The Chemistry and Manufacture of Powder Coating Resins

SUMMARY

Epoxy Resins

Epoxy resins are characterized by the epoxy group (epoxide), which occurs as the reactive end point of linear molecules. It is these reactive end groups with the hydroxyl

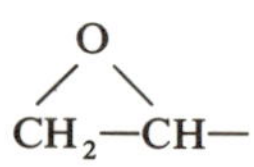

groups (OH) found in the linear chain which can undergo the wide variety of cross-linking reactions to give the exceptional adhesion, toughness and chemical resistance associated with epoxy resins.

The simplest epoxy compound of the type suitable for powder coatings is the diglycidyl ether of diphenylolpropane (bisphenol A). It is a low viscosity liquid of molecular weight 340 but in practice, the theoretical structure is difficult to obtain.

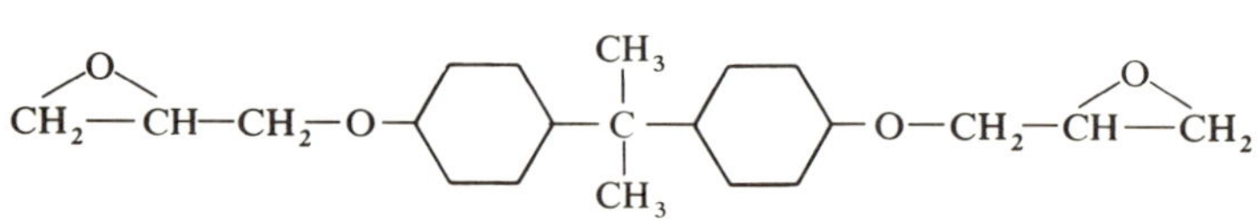

It is possible to increase the chain length of this basic epoxy compound by altering the ratio of the co-reactants and the reaction conditions, so that solid resins of higher molecular weight may be obtained.

A typical epoxy resin structure (made by the process shown on page 17) is shown in fig. 2 where *n* may lie between 4 and 9 for products used in powder coatings.

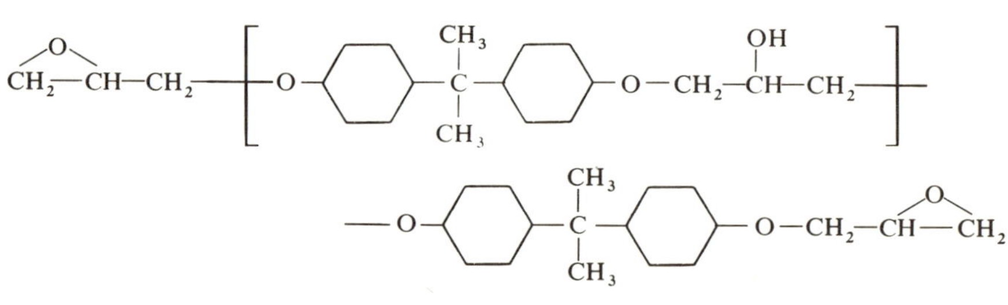

The ether linkages (—O—) of the epoxy resin confer the excellent chemical resistance while the high aromatic content is responsible for reduced resistance to chalking and poor light and heat stability. Its very good adhesive properties are effected by the small, highly polar, epoxy and hydroxyl groups.

Techniques currently used for the manufacture of epoxy based powder coatings require epoxy resins with melting points in the range 75–115°C with molecular weights lying between 800 and 2,000.

Epoxy resins are not normally specified by molecular weight, but rather in terms of the epoxide equivalent weight, that is the number of grams of resin containing one gram equivalent weight of epoxide

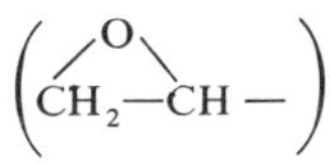

It will be seen from the tables of commercially available epoxy resins that epoxide equivalent weights of powder coating resins occur in the range 550 to 1,500.

The chemical reaction involved in the manufacture of epoxy resins is simple and, of short duration. It is, however, difficult to wash the resin free from salt formed during the reaction and the partial polymerisation of the resin to a higher molecular weight must be carefully controlled to ensure freedom from gelled particles which would mar the appearance of the powder coating.

Polyester Resins

This class of resins is possibly the most versatile for powder coating development. The resins are generally oil-free products formed by esterification of polybasic acids and polyhydric alcohols. Ester linkages do not possess the chemical inertness of the ether linkages present in the epoxy resins, but it is possible to obtain a high level of resistance to acid and alkaline conditions by the correct choice of reactants.

Polyester resin may, for cheapness, be based upon the readily available *o*-phthalic anhydride, but this substance is known to contribute to chalking and is more easily hydrolysed than other preferred polybasic acids.

Manufacture of polyester resins can be carried out in a reaction vessel suitable for the production of oil modified glyptal resins with the added requirement of a distillation column.

The cost and availability of the polyester raw materials are likely to be the major factors affecting formulation but there are still many possibilities available to the resin formulator.

Advantages to be gained by the use of polyester resins in place of epoxies are:—

Improved heat and light stability;

Better resistance to exterior weathering and freedom from chalking.

They offer greater versatility to the resin formulator and they are likely to be cheaper;

Good gloss and improved levelling properties may be obtained.

Disadvantages compared with epoxies are:—

The wider molecular weight spread which will lower the glass transition temperature and thereby cause "blocking" or "lumping" of the powder.

A tendency to hydrolyse which will result in coatings with lower chemical resistance.

Acrylic Resins

This class of compounds is obtained by the addition polymerization of unsaturated monomeric materials largely derived from the petrochemicals industry.

In recent months the world prices for such monomers have trebled or quadrupled, in line with most petrochemicals, and certain of the key components are now not readily available. Nevertheless, these resins are already well established in conventional paint finishes and widely used in the automotive and domestic appliance field.

Monomeric raw materials include the following compounds: Vinyl benzene (styrene); α-methyl styrene; the esters, hydroxy esters, glycidyl esters and amides of acrylic and methacrylic acids; acrylonitrile; and acrylic and methacrylic acids themselves.

Advantages which are generally claimed for acrylic resins are good light and heat stability, good detergent resistance, excellent physical properties of the cured film, freedom from chalking, good stain resistance and high gloss with good levelling characteristics.

The make-up of the acrylic resin will, however, greatly influence the physical and chemical properties of the cured film. A high proportion of styrene will adversely affect the light and heat stability and detract from chalk resistance, due to the presence of the aromatic ring. The excessive use of methyl methacrylate could promote brittleness of the cured film.

Crosslinking of the acrylic resin may be effected by five main reaction mechanisms, using internal or external co-reactants.

Table 2. Acrylic Functional Groups

Functional group of acrylic copolymer	Co-reactants and/or reaction mechanism
1. Hydroxyl. (—OH)	(a) Blocked isocyanate giving addition reaction with release of volatile blocking agent. (b) Alkoxy methylol melamine by a condensation process with release of water and formaldehyde.
2. Hydroxyl (—OH) and Carboxyl (—COOH)	Both groups contained in the copolymer so that the resin is self-cross-linked by a condensation process.
3. Amide. (—CONHCH$_2$OR) and hydroxyl (—OH)	Both groups contained in the copolymer and self cross-linked by a condensation process.
4. Carboxyl (—COOH)	This type is mainly cross-linked with epoxy resins by an addition mechanism resulting from cleavage of the epoxy ring.
5. Epoxy (—CH$_2$—CH—CH$_2$) \O/	Generally introduced to polymer by glycidyl methacrylate. The product is capable of cross-linking with a poly carboxylic acid (as 4 above). Reaction is by addition mechanism with ring cleavage.

Acrylic resins for use in conventional paint finishes are invariably manufactured by solution polymerization methods. These methods cannot be used for the production of acrylic resins suitable for powder coatings because of the difficulty of removing all traces of solvent, and it is also not possible to obtain the necessary high conversion rate of monomer to polymer which is required in a system free from volatile components. Precise control of molecular weight spread cannot be obtained by solution polymerization and this would adversely effect the "caking" characteristics of the finished acrylic powder.

Bulk polymerization methods are therefore employed and this involves the use of specialized reaction vessels for large scale manufacture and precise control of polymerization, which is often a two stage process.

Acrylic resins will become an essential part of powder coatings because they are so versatile. But it will be difficult for the small powder manufacturer to produce his own resins and it is this factor which will place the acrylic powder market in the hands of specialized resin manufacturers.

This restriction will, however, renew activity in the development of polyester resins which can be produced by the small manufacturer and so the market is unlikely to be dominated by acrylics.

EPOXY RESINS

The raw materials used in the synthesis of epoxy resins are epichlorohydrin and bisphenol A (diphenylol propane).

Epichlorohydrin

Epichlorohydrin is a colourless, mobile liquid with an irritating odour, its formula is:

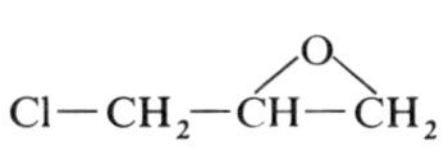

It is produced by the chlorination of propylene and is manufactured as a stage in the production of synthetic glycerol. There are few independent manufacturers of epichlorohydrin since it is a by-product of the petroleum and natural gas industries. The major producers are also well established manufacturers of epoxy resins.

Bisphenol A

Bisphenol A or bis (4-hydroxyphenyl) dimethylmethane is a flaked solid of melting point 153°C its formula is:

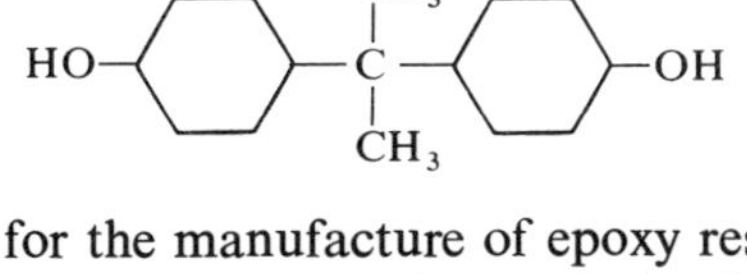

The direct method for the manufacture of epoxy resins is based upon the reaction of epichlorohydrin and bisphenol A in the presence of excess caustic soda. Synthesis of high molecular weight resins is a two stage reaction involving dehydrohalogenation of the epichlorohydrin followed by the addition of further bisphenol A to the epoxy group.

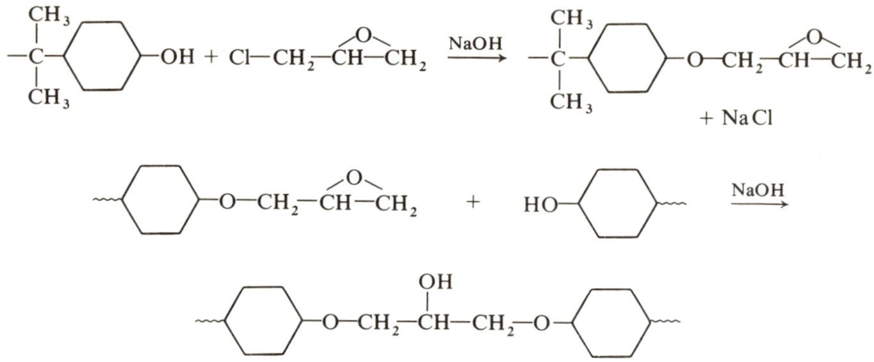

The degree of polymerization is therefore largely dependent upon the ratio of epichlorohydrin to bisphenol A using a slight excess of caustic over the amount necessary to neutralize the hydrochloric acid. If the excess amount of caustic is not carefully controlled it can catalyse the polymerization by the epoxy groups producing fractions of very high molecular weight. In the early years of epoxy resin manufacture the presence of gelled particles in the resin was particularly troublesome to the coatings industry, especially in the field of epoxy ester production.

The effect of varying the ratio of reactants is shown in Table 3.

Table 3. The Effect of Varying Epoxy Resin Components

Epichlorohydrin/ bisphenol A mol. ratio	Caustic Soda/ Epichlorohydrin mol. ratio	Melting point °C	Molecular weight	Epoxide equivalent
1·5	1·25	75	736	550
1·3	1·29	95	1,050	820
1·2	1·29	110	1,350	1,140

To manufacture epoxy resins by the above process it is necessary to use a heavy duty reaction vessel capable of operating at high pressure to assist separation of the resin/water layers and the subsequent high temperature washing to remove the salts formed in the reaction.

The bisphenol A is dissolved in a solution of caustic soda and the epichlorohydrin is added at 50°C. After the initial exotherm, the temperature is raised above 100°C and a pressure of 25–30 atmospheres is applied. Washing is effected by circulating hot water through the vessel, while maintaining the pressure, for several hours. The resin is finally separated from the water layer and dried at temperatures of 135–140°C.

Criticisms of the above method are:

(a) The difficulty of washing the resin to remove salt and traces of caustic completely.

(b) Poor control of molecular weight distribution due to side reactions promoted by the presence of alkali and water.

Most of the commercially available epoxy resins are today manufactured by the advancement process. This is a two stage operation involving the formation of a low viscosity resin of molecular weight between 750 and 900. It is easier to wash and purify this product so that it may be further combined with a calculated amount of bisphenol A and a polymerization catalyst by an addition reaction between the terminal epoxide groups and the phenolic hydroxyl.

Epoxy based powder coatings are usually manufactured from resins with epoxy equivalents between 500 and 1,000. This range is defined by the softening points of the resins; resins with epoxy equivalents lower than 500 have melting points in the region of 60–70°C and produce powder coatings with a pronounced tendency to cake. Resins with epoxy equivalents greater than 1,000 have softening points above 100°C, producing powders with poor flow properties. The most suitable epoxy resins for powder coatings will be found in the range 850–950 epoxide equivalent.

Correctly formulated epoxy powder coatings will depend as much upon the choice of resin as they will on the choice of hardener. The role of the hardener will be discussed fully in Chapter 3.

A compromise in the choice of resins can assist the development of optimum physical and chemical properties in the cured coating. At low epoxide

Table 4. Commercially Available Epoxy Resins

Reference	Supplier	Type	Epoxide equivalent	Softening point °C[a]	Viscosity c.p.s.[b]
Araldite 7004	Ciba-Geigy Ltd.	Unmodified epoxy	715	72	490
Araldite 7084	Ciba-Geigy Ltd.	Unmodified epoxy	905	95	500
Araldite 2808	Ciby-Geigy Ltd.	Modified epoxy	670	72	415
DER 662U	Dow	Unmodified epoxy	635	85	240
DER 663U	Dow	Unmodified epoxy	785	93	425
DER 664U	Dow	Unmodified epoxy	925	100	670
DER 642U	Dow	Unmodified epoxy	540	90	425
XD.5559	Dow	Partial ester	915	93	540
Epikote 1055	Shell Chemical	Unmodified epoxy	850	94	450
Epikote 2057	Shell Chemical	Unmodified epoxy	890	98	570

[a] Determined by Durrans Mercury Method.
[b] Determined as a 40% solution in butyl carbitol at 25°C.

equivalent the resins when cured will show improved solvent, chemical and boiling water resistance with good flow characteristics. High epoxide equivalents will give improved flexibility and impact resistance and better storage stability to the powder. Hardness and adhesion will not be significantly different within the quoted range of epoxide equivalents.

Flexibility, water resistance and flow properties of epoxy powder coatings may also be improved by reacting epoxy resins with dimerized acids to produce partial esters.

Table 4 lists the comparative physical properties of epoxy resins recommended for use in powder coatings by the major suppliers of resins in Europe.

The manufacturers literature for epoxy resins describes a range of values for each of the three properties. An average figure has been quoted in the tables to allow for future adjustment of the limits as manufacturing techniques are improved.

POLYESTER RESINS

The major chemical reactions occurring during the formation of polyesters are condensation reactions of esterification between polybasic acids and polyhydric alcohols; etherification of the polyols, and ester interchange.

A simple esterification reaction is an equilibrium system where the reaction is reversible.

$$R'.COOH + R.OH \rightleftharpoons R'.COOR + H_2O$$

When the reactants both possess more than one reactive group then polyesterification can result from a stepwise condensation process by removal of the water formed during reaction. An example is the formation of a linear polymer by reacting ethylene glycol and adipic acid.

$$n(HOCH_2CH_2OH + HOOC(CH_2)_4COOH)$$

$$\rightleftharpoons HO\left[(CH_2)_2 . OOC . (CH_2)_4 . COO\right]_n H + (n-1)H_2O.$$

With increasing molecular weight the product changes from a mobile liquid to a solid of high melting point.

Linear polymers will be formed if the acid and alcohol each have only two reactive or functional groups. If reactants are introduced with more than two functional groups then chain branching of the basic resin will occur with further crosslinking between resin molecules. The degree of cross-linking must be carefully controlled to prevent gelation of the resin to an insoluble and infusible product.

In the design of polyester resins for powder coatings it is desirable to keep chain branching to a minimum so that the flow of the powder coating is not affected by the molecular complexity of the resin.

However, a thermoset polyester powder coating must be cross-linked during the stoving process to render the film infusible. To complete the cross-linking reaction during stoving it will be necessary to incorporate certain curing agents in the compounding process and these are described in the next chapter. It is possible to introduce these curing agents during the resin manufacture but experience has shown that even latent curing agents do adversely affect the storage stability of the resin blends.

The requirements of a suitable powder coating polyester are that it shall possess a melting range similar to that described for the epoxy resins (85–100°C) and a glass transition temperature above 65°C. The acid value should be below 3 and the hydroxyl value in the range 115–130 since crosslinking will invariably take place by further reaction with the excess hydroxyl groups.

Polyesters are difficult to produce because of the wide molecular weight spread within the resin. The average molecular weight of 3000–5000 may contain undesirable amounts of resin with molecular weight of 10,000 and above, thereby affecting the flow, compatibility and wetting properties; and, even less desirable are the low molecular weight fractions which adversely affect the glass transition point and cause blocking and lumping of the finished powder.

It is possible to minimize the formation of higher molecular weight material by close control of manufacture and to some extent the low molecular weight fractions result from cyclization and steric hindrance due to the directional influence of the reactive groups and their efficiency of reaction rate. For example, it is known that phthalic anhydride can react to form closed rings of low molecular weight resin due to the close proximity of the two reactive carboxyl groups, and glycerine does not react smoothly because the secondary hydroxyl group reacts at a lower rate than the two primary hydroxyls. Unfortunately, many of the polybasic acids which do not promote closed ring structures will only react at elevated temperatures and these higher temperatures allow etherification between hydroxyl groups and ester interchange resulting in the formation of high molecular weight components.

Resin formulating possibilities are considerable and the choice of reactants will be dictated by local availability, cost and the properties desired in the final resin.

A list of common reactants is given and the properties they impart to the polyester are detailed to assist the formulator to choose the correct combination of reactants.

Raw Materials

POLYBASIC ACIDS

Phthalic Anhydride

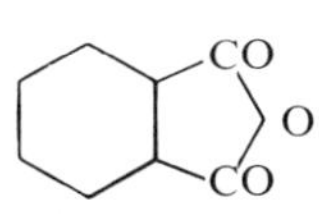

This is the cheapest and most readily available dibasic acid. The acid sublimes during reaction and promotes low molecular weight fractions. Esters formed from this acid are fairly readily hydrolysed and its use therefore imparts to resins poor water resistance and durability characteristics. Inferior chalk resistance properties result from the excessive use of phthalic anhydride.

Isophthalic Acid 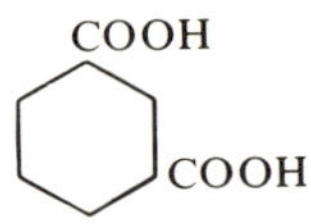

This isomer of orthophthalic acid does not form the anhydride. It has a higher melting point and therefore requires a higher reaction temperature. Esters including this acid are characterized by greater resistance to hydrolysis with corresponding improvement in water resistance and exterior durability. It is less prone to chalking than phthalic anhydride but is not completely free from this fault. Flexibility and film toughness are improved but the ester often possesses a darker colour due to the higher reaction temperature.

Maleic Anhydride 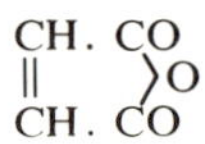

This dibasic acid has a functionality of four due to the presence of the double bond and it is therefore extremely reactive if used in large quantities. The inclusion of 3–5% with phthalic anhydride will improve the reactivity of the polyester and enhance water resistance properties. It is non-aromatic and therefore enhances colour stability. Before it is included in the polyester formulation it may be converted to a dibasic acid with a functionality of two by Diels–Alder reaction with cyclopentadiene

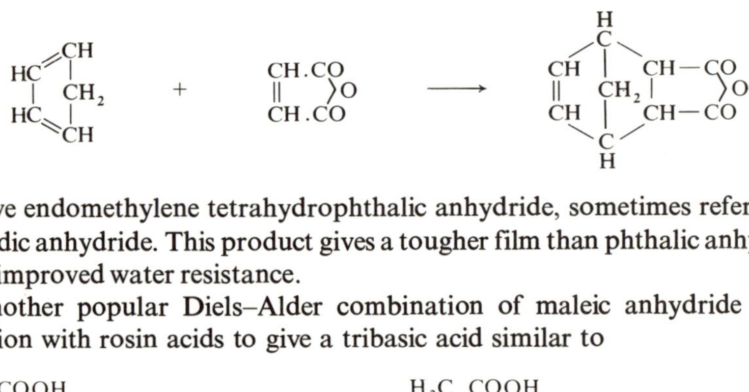

to give endomethylene tetrahydrophthalic anhydride, sometimes referred to as nadic anhydride. This product gives a tougher film than phthalic anhydride with improved water resistance.

Another popular Diels–Alder combination of maleic anhydride is the reaction with rosin acids to give a tribasic acid similar to

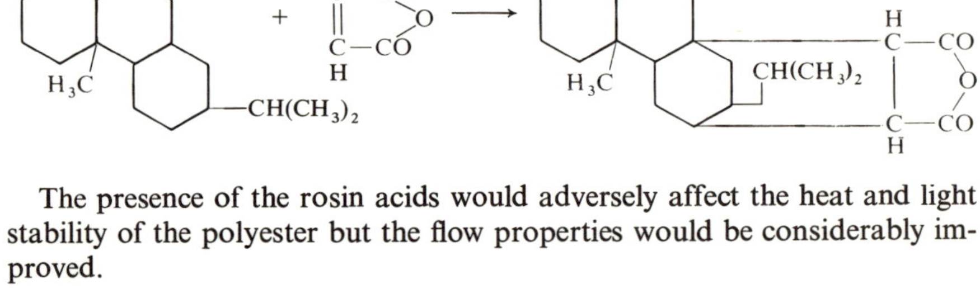

The presence of the rosin acids would adversely affect the heat and light stability of the polyester but the flow properties would be considerably improved.

Adipic Acid $HOOC(CH_2)_4COOH$

Widely used in the manufacture of nylon, this dibasic acid imparts excellent flexibility properties to the polyester due to its linear chain structure. It does not promote chalking and is therefore recommended for inclusion in exterior quality polyesters. There are other commercially available dibasic acids in this series which may be used to impart flexibility, including azelaic acid $HOOC(CH_2)_7COOH$ and sebacic acid $HOOC(CH_2)_8COOH$. Generally acids with an even number of CH_2 groups are preferred.

Trimellitic Anhydride

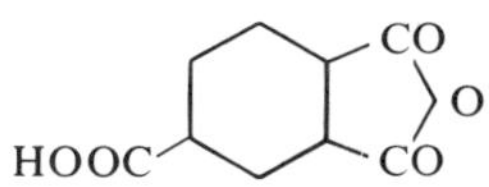

The trifunctionality of TMA gives high reactivity and branch chaining. It is best used as a partial replacement for difunctional acids to improve chemical and water resistance of polyesters. Excellent heat and light stability is imparted by TMA and it has low volatility.

Trimellitic anhydride is readily hydrolyzed to trimellitic acid, a transition that can occur during storage at room temperature.

Dimethyl terephthalate

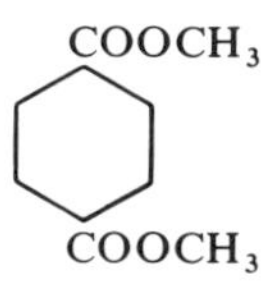

Terephthalic acid, the para isomer of orthophthalic acid is almost impossible to use in polyester manufacture due to its high melting point and poor solubility. The dimethyl ester, however, may be used in small quantities to react through ester interchange. Good flexibility, resistance to hydrolysis and thereby improved chemical resistance would result from the inclusion of this ester but its high cost is prohibitive.

POLYHYDRIC ALCOHOLS

Ethylene Glycol $HO{-}CH_2{-}CH_2{-}OH$

Diethylene Glycol $HO{-}CH_2{-}CH_2{-}O{-}CH_2{-}CH_2{-}OH$

While it is possible to incorporate small amounts of these two glycols, their use in powder coating polyesters is not recommended because of their hygroscopic nature and the adverse effect that traces of unreacted glycol could have upon the water resistance of the cured film. They also tend to cause blocking of the powder coating.

Neopentyl Glycol

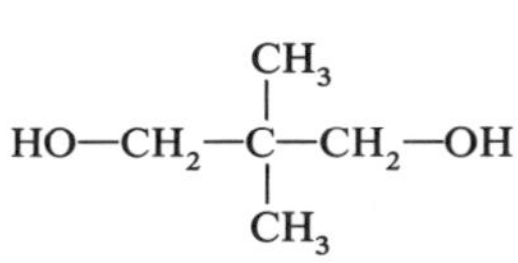

Trimethyl pentanediol

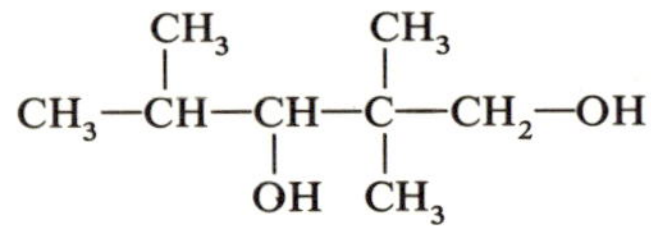

Cyclohexanedimethanol

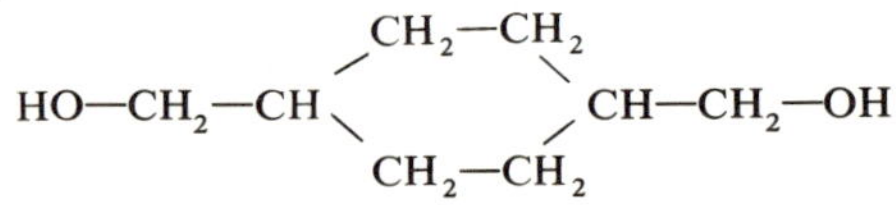

The three glycols NPG, TMPD and CHDM are recommended for use in powder coating polyesters[1] to impart good heat stability, stain resistance, flexibility and pencil hardness when the polyester is cross-linked with amino compounds or isocyanates.

Glycerol

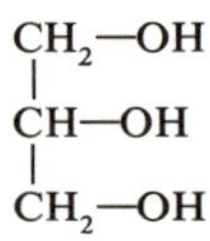

This polyol is included for the sake of completeness but is not recommended as a component of powder coating polyester resins due to its uneven rate of reaction at the secondary hydroxyl group; this promotes a wide molecular weight distribution.

Trimethylolethane

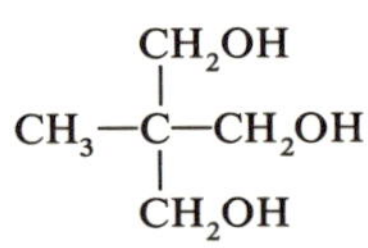

TME reacts smoothly in polyester resin formation due to its three primary hydroxyl groups. It should only be used as a partial replacement for glycols for its high functionality will result in branch chaining and high molecular complexity. Improved alkali and water resistance will result from the inclusion of TME.

Trimethylolpropane

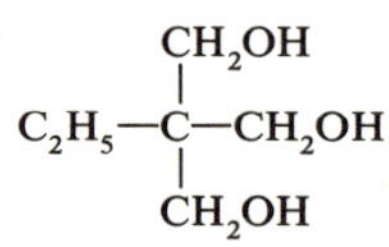

Properties of this polyol are similar to those of TME but it is less reactive. Polyesters based on TMP will be more compatible with other resins than those employing an equivalent amount of TME.

Formulating Principles

The objective of the formulator is to produce a polyester resin of average molecular weight in the region of 2,000–3,000, with a melting range between

85°C and 105°C, and a glass transition temperature above 65°C. The acid value should lie in the range 3–5 with a hydroxyl value of 130–150.

Stoichiometric amounts of acids and alcohols will not react to the low acid values required and gelation of the product is the most likely result if this is attempted.

In practice it is necessary to use an excess of hydroxyl over acid. Powder coating polyesters generally require an excess of hydroxyl of between 15 and 25%. Increasing amounts of excess hydroxyl will improve the compatibility of the polyester with other resins but they may lower the glass transition temperature.

An optimum value for excess hydroxyl is close to 20%.

It has already been mentioned that an overall functionality of two will result in the formation of linear polyesters while increasing this figure will lead to chain branching and eventually to gelation.

Some measure of chain branching is desirable to enhance the chemical resistance qualities and it is suggested that overall functionalities of 2·05 to 2·1 should be employed.

Typical starting point formulations are given in publications issued by Eastman Chemical Company[(1)] and AMOCO.[(2)] The formulator will also find the general principles fully explained in the text book by Patton.[(3)]

ACRYLIC RESINS

In this review of acrylic resins suitable for use in powder coatings, it is likely that many of the methods suggested are covered to some extent by existing patent literature. The considerable worldwide interest in these resins has resulted in many new and novel approaches to the problem of close molecular weight control of powder coating acrylic resins.

The coatings chemist will be familiar with the techniques of solution polymerization of vinyl and acrylic monomers to produce solvent soluble acrylic resins for conventional paint finishes, but it is unlikely that the bulk polymerization method will have been examined as a procedure for the manufacture of coating resins, due to the inherent production difficulties.

A powder coating acrylic resin must possess a solids content close to 100%, a molecular weight distribution between 3,000 and 5,000, a glass transition temperature above 65°C and a melting range between 75 and 105°C. The stipulation of high solids content favours the bulk polymerization process.

A major drawback of the bulk polymerization method is the highly exothermic reaction occurring in large commercial batches and the dissipation of this excess heat becomes considerably more difficult as the resin viscosity increases. Unless really effective stirring can remove localized heat pockets there is a distinct danger that an explosion can result.

To reduce the danger of serious exotherms in large scale production, the reaction screw technique for continuous production of acrylic resins has been developed. The relatively small volume of the screw facilitates the removal of excess heat, but the products manufactured by this process are invariably of molecular weight higher than 10,000.

Tubular reactors can be used for bulk polymerization processes since they offer a larger cooling area, but the basic problem of the continuous method is the close control of molecular weight distribution.

The method which allows best control of molecular weight and melting range is the batch process, and the conditions of reaction must be designed to control the rate of reaction.

Organic monomers can be bulk polymerized at elevated temperatures to give a medium viscosity syrup which can be polymerized further at low temperatures.

A preferred method, however, is the partial polymerization of the monomer, at temperatures in the region of 100°C, by the continuous addition of an initiator which decomposes into free radicals below 100°C.

A conversion of monomer of between 70 and 80% can be achieved with controlled dissipation of the heat of reaction. The efficiency of this fast radical initiation falls off rapidly above 70% conversion; it is therefore necessary to raise the temperature to above 130°C and complete the conversion by use of an initiator which decomposes into free radicals only above 100°C.

When a conversion greater than 95% is attained, the remaining free monomer can be removed by vacuum distillation or nitrogen purging.

In the first stage of this two step polymerization there is no advantage to be gained by the continued addition of fast forming, free radical initiators since an excess will not promote the reaction velocity and this excess can have an adverse effect upon the final softening point of the resin, for it could act as a plasticizer.

It is unsafe to attempt to carry out the bulk polymerization as a one-step process, using only an initiator decomposing above 100°C, for the free radical formation can be much slower than the rate of initiator addition. A sudden increase in heat could cause this excess latent initiator to form free radicals more quickly and promote a reaction of explosive violence.

The control of molecular weight may be effected by the addition of small quantities of chain terminators or regulators. For this purpose commonly used substances are mercaptans, chlorinated compounds and a wide range of organic solvents.

First stage polymerization by fast forming free radical initiators is best controlled by the continuous addition of the initiator dissolved in a portion of the monomer. In the second stage of polymerization the initiator may be added in several portions.

Choice of Monomers

Polymerization will occur by addition mechanism between the olefinic unsaturated bonds of the different co-monomers. Crosslinking, by reaction with external curing agents or resins, will take place at the pendant polar groups and these may include epoxy, hydroxyl, carboxyl or amide.

The backbone skeletal structure of the polymer will be provided by monomers which do not contain pendant polar groups. Proportions of monomer in the region of 70% will usually be provided by this relatively simple class of compounds.

They include:

Styrene (vinyl benzene) $CH_2{=}CH$ 

Methyl methacrylate $CH_2{=}\underset{\displaystyle CH_3}{\underset{|}{C}}\,.\,COOCH_3$

Ethyl acrylate $CH_2{=}CH\,.\,COOC_2H_5$

A high proportion of styrene will result in a polymer with poor light and heat stability and the exterior durability will also be poor. The presence of the aromatic group is responsible for these poor performance characteristics.

Methyl methacrylate will polymerize to a water white resin with good heat and light resistance properties.

It is, however, rather brittle and would give mechanical properties below the standards required for surface coating applications.

Ethyl acrylate will also polymerize to give a resin with good heat and light resistance but excessive amounts of this monomer will drastically reduce the hardness of the coating. It has a characteristic pungent odour which is difficult to remove from the polymer.

More costly acrylic and methacrylic esters may be used but it is doubtful whether the improvement in film performance of the copolymer justifies the additional expense.

The carboxyl group is introduced by the use of up to 20% of acrylic acid $CH_2{=}CH\,.\,COOH$ or methacrylic acid $CH_2{=}\underset{\displaystyle CH_3}{\underset{|}{C}}\,.\,COOH$.

Introduction of the hydroxyl group is achieved by incorporating quantities of hydroxy alkyl acrylates or methacrylates up to 50% of the total weight of co-monomers. There are numerous compounds of this class but those most commonly available include

Hydroxyethyl acrylate $CH_2{=}CH\,.\,COOCH_2CH_2OH$

Hydroxy propyl acrylate $CH_2{=}CH\,.\,COOCH_2\underset{\displaystyle OH}{\underset{|}{CH}}\,.\,CH_3$

and the corresponding methacrylate derivatives.

Hydroxyethyl methacrylate $CH_2{=}\underset{\displaystyle CH_3}{\underset{|}{C}}\,.\,COOCH_2CH_2OH$

Hydroxypropyl methacrylate

Other co-monomers that may be used to incorporate polar groups include:

Acrylamide $CH_2{=}CH\,.\,CONH_2$

or amide derivatives with the following general structure:

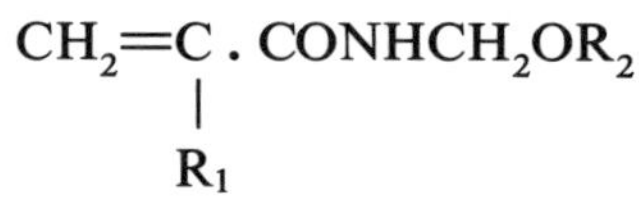

The epoxy group may be introduced by including glycidyl methacrylate in the blend of co-monomers

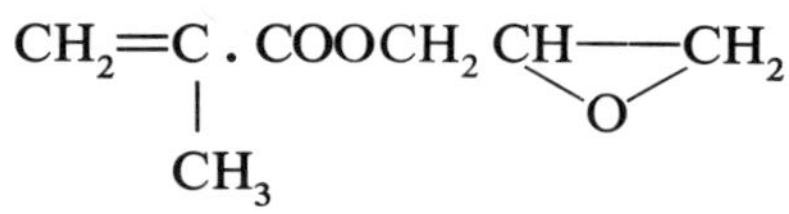

These various polar groups may be used singly, to provide one chemical type to be crosslinked by reactions with an external curing agent, or it may be possible to include two reactive polar groups of different chemical type so that the polymer may be self crosslinking by internal condensation or re-arrangement reactions.

Some of the possible self-cross-linking polar groups are:

OH	and	COOH
$CONHCH_2OR$	and	OH
$COONHCH_2OR$	and	OH

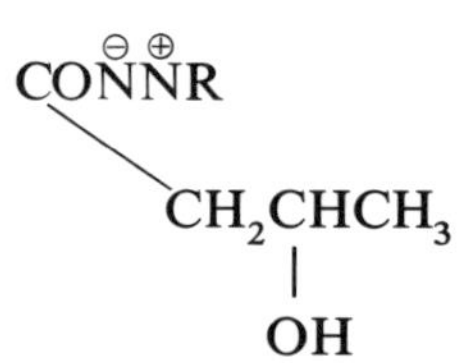

and OH

External crosslinking mechanisms will be described in the next chapter.

Choice of Initiators

Fast reacting initiators which decompose at temperatures below 100°C include: benzoyl peroxide, lauroyl peroxide, cyclohexanone hydroperoxide, tertiary butyl perpivalate and percarbonates such as di-isopropyl peroxy dicarbonate, nitrogen compounds represented by azodiisobutyrodinitrile and certain sulphonyl peroxides.

Slowly decomposing initiators include: tertiary butyl perbenzoate, ditertiary-butyl peroxide, cumyl hydroperoxide and methyl ethyl ketone peroxide.

MISCELLANEOUS POWDER COATING RESINS

While the attention of the powder coatings formulator is currently directed to epoxy, polyester and acrylic finishes, the resin chemists will be aware that there are also possibilities to develop amino resins and it is known that one resin manufacturer has already introduced a melamine resin for cross-linking powder coating polyesters.

Phenolic resins are used in the surface coatings industry as coatings in their own right, and they are also widely used in conjunction with epoxide as can and drum linings.

Addition reactions are most favoured in powder coatings technology for they ensure the concept of 100% non-volatile coatings. The usefulness of condensation mechanisms will become more apparent with the advent of thin film techniques.

A modified vinyl chloride resin has recently been designed for use in powder coatings with excellent adhesion to unprimed metal surfaces and this has aroused much interest because of the excellent weathering properties of vinyls.[4]

Polyurethane powder coatings are normally produced by the addition of a blocked isocyanate to a hydroxyl containing resin. The most common types are polyesters or polyethers. It is possible however partially to crosslink the base resin with a blocked isocyanate and this modification is already mentioned in patent literature.

REFERENCES

1. Eastman Industrial Chemicals. Publication No. 168.
2. A.M.O.C.O. Bulletins TMA 62b and DT 2a.
3. T. C. Patton. *Alkyd Resin Technology*, Interscience Publishers Inc., New York (1962).
4. Union Carbide. Product E 2000.

3

The Cross-linking Reactions of Thermoset Powder-Coating Resins

SUMMARY

The agents used for crosslinking or curing in the thermosetting reactions of powder coatings, fall into two main classes:

(a) catalytic curing agents which serve as initiators for homopolymerization of the resin molecules and
(b) polyfunctional curing agents which serve as co-reactants and therefore become an integral part of the cured film.

Curing agents must be chosen carefully to conform to the following requirements:

1. Compatibility with the base resin.
2. To minimize deterioration of the mixed powder during storage, agents should be latent at room temperature.
3. They themselves must be stable compounds at ambient temperatures and not greatly affected by the atmospheric moisture content.
4. Ideally they should remain latent during the compounding operation and only become effective as curing agents at temperatures in the region of 150°C. Possession of this quality will markedly affect the flow properties of the powder since it is essential that the applied powder be allowed to flow, at its melting point, into a smooth film and "wet" the substratum before the curing reaction begins.
5. They should not present a toxic hazard, neither during compounding nor in the final product.
6. For ease of incorporation, they should be solids at ambient temperature and finely ground to assist homogeneous dispersion at the premixing stage.
7. They should not produce volatile by-products during cure.

Epoxy resins are now (mid 1970's) the most widely used resins for powder coatings The actual curing reactions are ionic in nature and both anionic and cationic polymerizations can occur.

First generation powder coatings mainly used dicyandiamide as the curing agent, a substance which required high stoving temperatures to effect curing. In the second generation of epoxy powders accelerated forms of dicyandiamide or substituted versions of it are used. Acid anhydrides may also be used to cure epoxy and polyester resins but they are not entirely compatible at the high concentrations needed to effect cure.

The level of gloss in the stoved film is dependent on the amount and type of curing agent.

CURING OF EPOXY RESINS

Crosslinking of the epoxy resin is achieved through epoxide or hydroxyl groups. Curing is normally an addition reaction without the formation of byproducts, the reactions being exothermic and with their reaction rates increasing with increasing temperature.

Steric hindrance factors can greatly affect the ease of reaction.

Reaction with Primary Amines

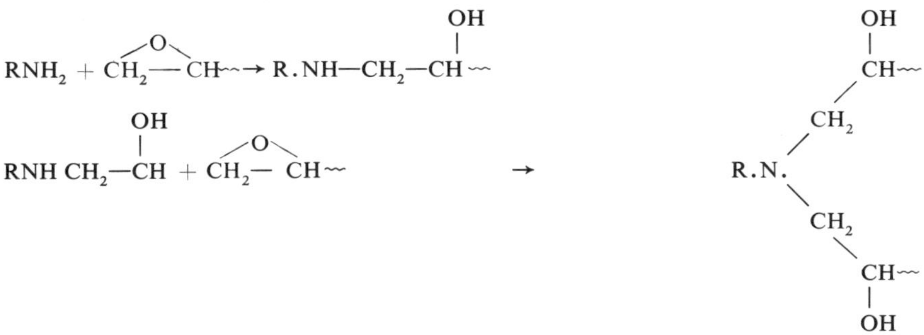

Reaction with Secondary Amines

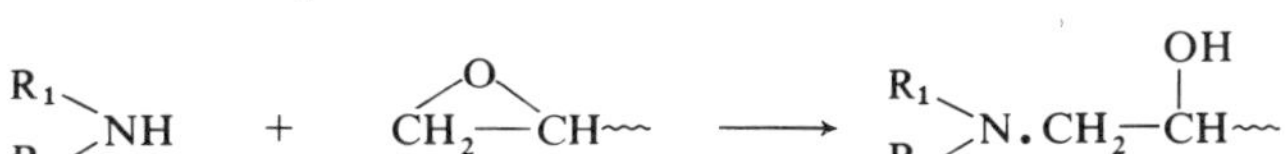

This reaction does not require a catalyst although it has been found that small amounts of hydroxyl containing materials (phenol, water, isopropanol) can accelerate the reaction. It is generally considered that the hydroxyl group assists the opening of the epoxide ring in the transition state by hydrogen bonding to oxygen.

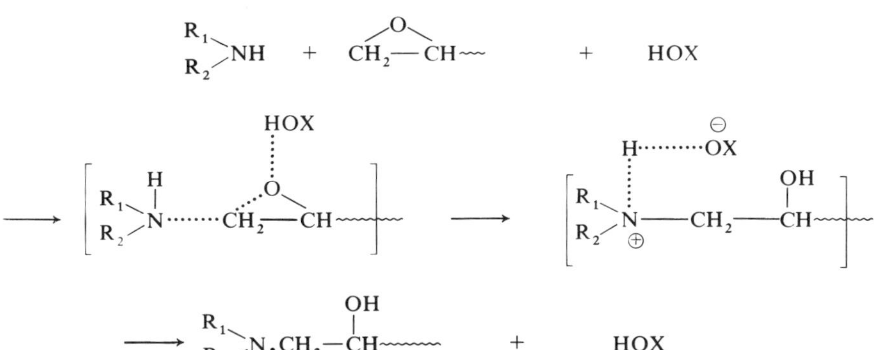

Hydrogen donors, which can serve as reaction accelerators, are:

$$—OH, —COOH, —SO_3H, —CONH_2$$

Hydrogen acceptors, which act as retarders, are:

$$—OR, —COOR, —CO—, —CN, —NO_2$$

Aromatic primary and secondary polyamines have lower basicity than the aliphatic amines and it is possible to react them with epoxy resins, at relatively low temperatures, so that only a portion of the epoxide groups are combined. This partial reaction is known as B stage curing, an effect which does not occur with aliphatic amines.

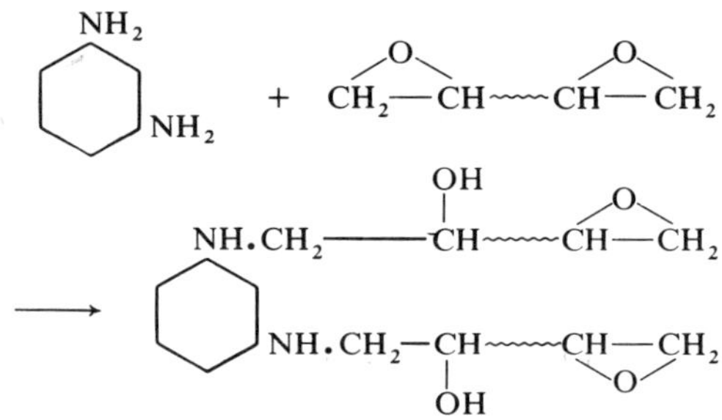

CATALYTIC CURING OF EPOXY RESINS

There are two main classes of compounds—bases and acids—that are employed in the catalytic crosslinking of epoxy resins which takes place only through the epoxide ring.

1. Lewis bases are substances which possess an unshared pair of electrons in their outer orbit and it is by donating these electrons that nucleophilic catalysis can be brought about. Tertiary amines are such bases and they are commonly used as catalysts in conjunction with primary and secondary amines or acid anhydrides; they are often incorporated in powder formulations to act as accelerators in powder curing reactions. Imidazoline derivatives may be used alone as nucleophilic catalysts.

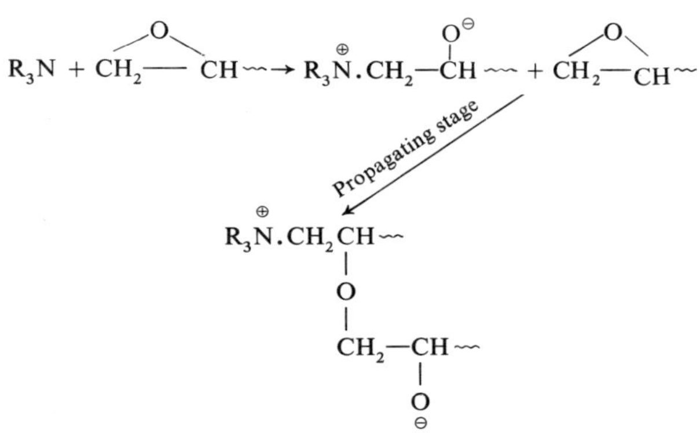

Alkali metal hydroxides will also behave in this manner which underlines the importance of completely removing traces of alkali during the manufacture of epoxy resins.

2. Lewis acids have an electron deficiency in their outer orbits and can accept electrons from another atom to form a chemical bond. The catalytic process is therefore electrophilic.

Boron trifluoride complexes are the best known examples in this class of catalyst.

This complex then reacts further with the hydroxyl group.

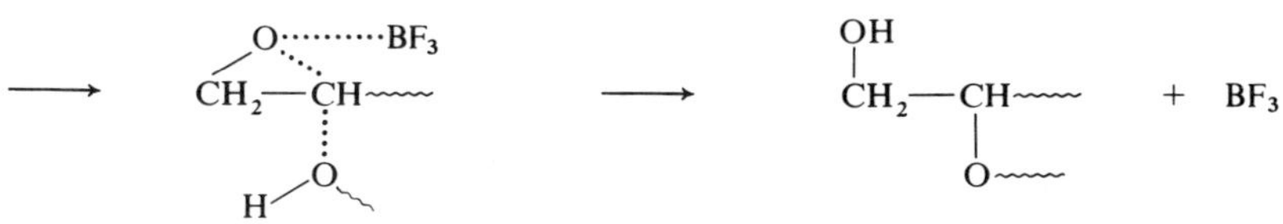

Since BF_3 is a highly corrosive gaseous compound it is normally used as the amine or ether complex e.g. $F_3B.NHR$.

REACTION WITH ACID ANHYDRIDES

Polybasic acids are not recommended as curing agents for powder coatings because the formation of volatile by-products must be avoided.

The first stage of the reaction with acid anhydrides is the opening of the anhydride ring by one of the hydroxyl groups in the epoxy resin with the formation of a half ester which then reacts further with the epoxide ring to form a hydroxy-diester.

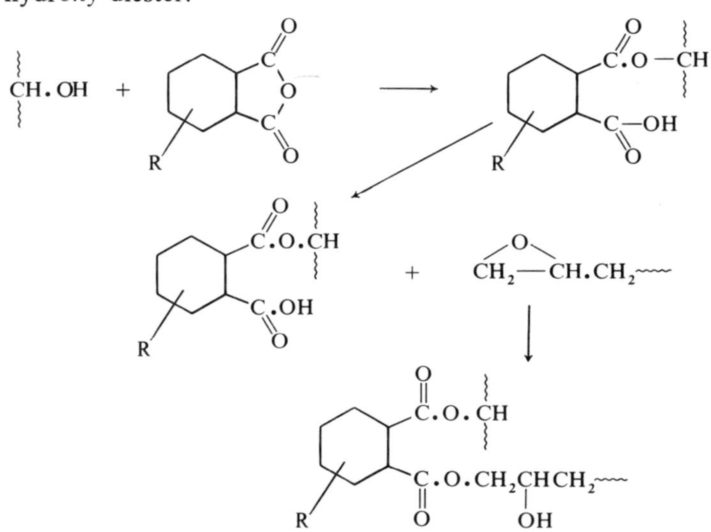

Tertiary amines may also be used to open the anhydride ring.

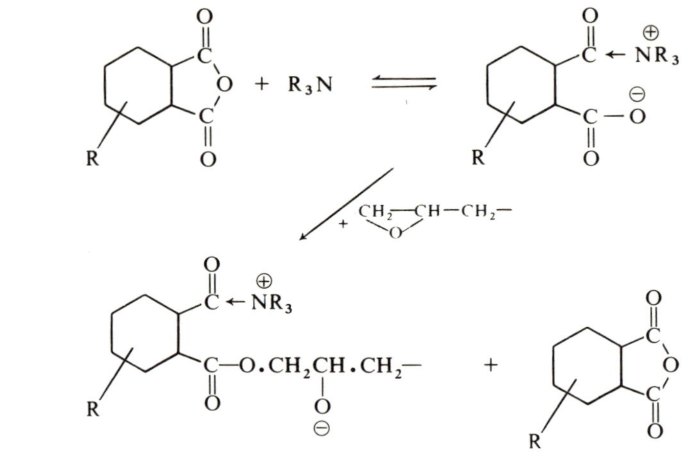

15

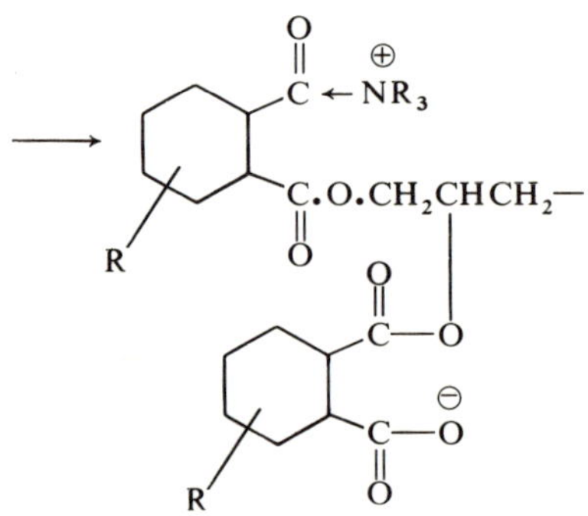

Boron trifluoride complexes may also be employed to open the anhydride ring.

$$RNH{:}\ BF_3 \rightleftharpoons \overset{\oplus}{H} + \overset{\ominus}{RN}{:}\ BF_3$$

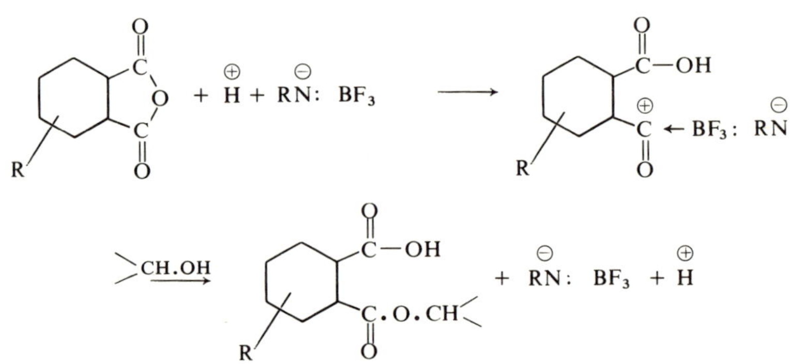

The important stage of the reaction of epoxy resins with acid anhydride is the opening of the anhydride ring. It has been shown that this may be achieved in several ways:

(a) The hydroxyl group in the resin molecule employed without a catalyst will result in esterification and etherification processes, while
(b) Tertiary amines will favour esterification and
(c) Boron trifluoride complexes will promote etherification.

It is generally undesirable to promote etherification since this will leave unreacted anhydride present in the cured film.

Catalysts which have been suggested for use with anhydrides include:

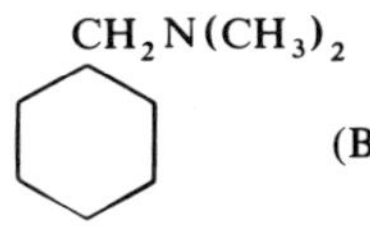

(Benzyldimethylamine)

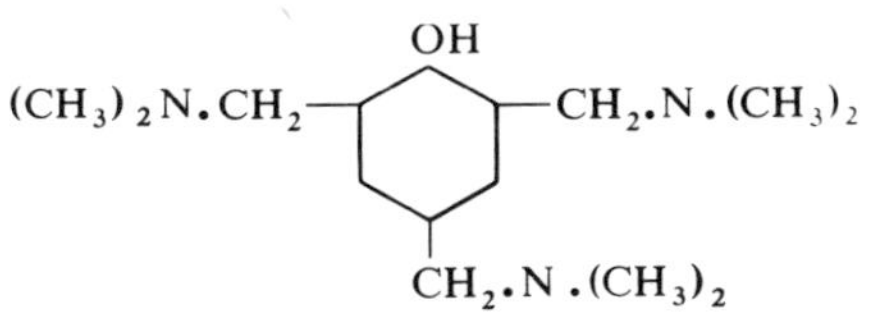

Tris(dimethylaminomethyl)phenol

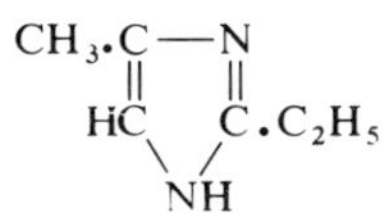

2 ethyl 4 methylimidazole

The accelerators are normally employed at 0·5–2·5 parts/hundred of resin.

DICYANDIAMIDE CURING AGENTS

This curing agent is most widely used in powder coating epoxy resin cross-linking and it is seen that the mechanism is that of an amine curing agent. The poor solubility of dicyandiamide and its high melting point are the main reasons for its poor curing rate. Substituted dicyandiamide derivatives are now freely available and these afford improved compatibility and reduced melting points.

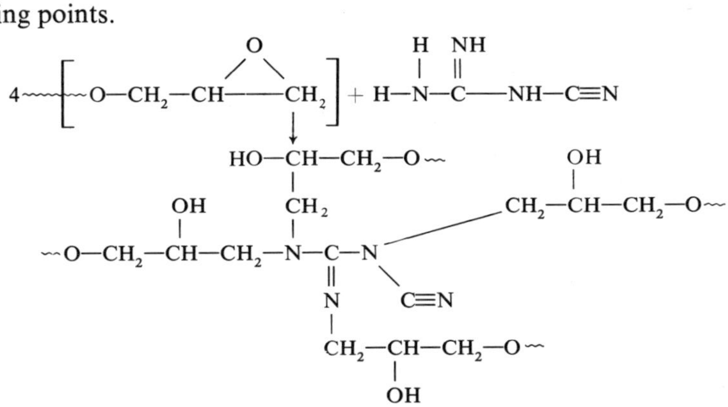

Substituted dicyandiamides have the general formula

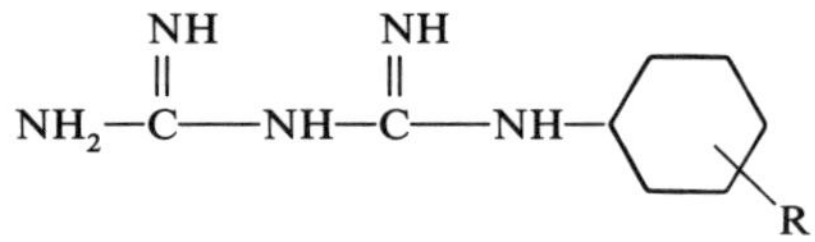

POLYAMIDE CURING AGENTS

This class of curing agents is best described as polyaminoamides produced by the reaction of polyamines with dimerized and trimerized vegetable oils.

$$\text{Polymerized fatty acid } -\overset{\displaystyle O}{\overset{\|}{C}}-\overset{\displaystyle H}{\overset{|}{N}}=\overset{\displaystyle H}{\overset{|}{N}}=NH_2$$

The members exhibit good compatibility with epoxy resins, low volatility and provide excellent flexibility properties in the cured film.

POLYURETHANE CURING AGENTS

Polyurethanes are characterized by a repeating urethane group in the polymer chain. This urethane group

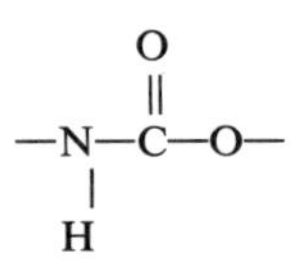

is usually obtained by addition polymerization of a di-isocyanate with a polymeric compound having pendant hydroxyl groups i.e.: epoxy resins, polyesters, polyethers, hydroxy acrylic resins.

It is thought that reaction occurs because of an electron deficiency on the carbon atom of the isocyanate group giving an electrophilic attraction to the nucleophilic centre of the hydroxyl group.

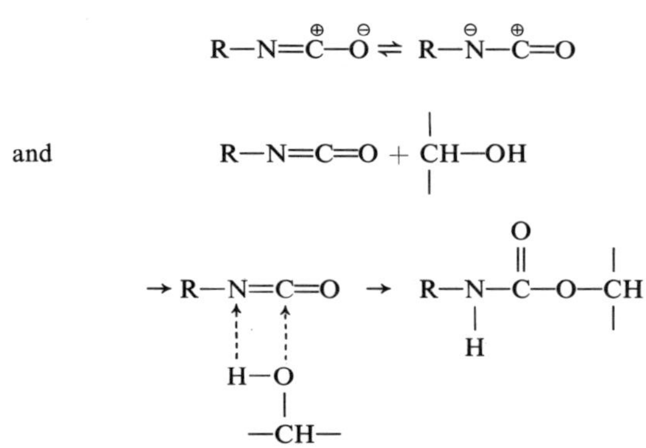

Isocyanates can only be used as cross-linking agents for powder coating resins if they are blocked to prevent reaction occurring at low temperatures. Blocking may be achieved with a variety of monohydroxy compounds many of which feature in patent literature and include the ∑-caprolactams[1] and the oximes[2].

The reactive NCO groups of the blocked isocyanate may be freed for crosslinking with the hydroxyl containing polymers, by heating to a temperature in the region of 150–200°C. Blocking agents are generally volatile at these temperatures and result in considerable loss of coating weight.

Aromatic diisocyanates unblock more readily than the aliphatic diisocyanates, but aromatic compounds generally lower the resistance of the final cured film to the discolouring effects of ultraviolet light.

The most readily available and economic of the diisocyanates are the low molecular weight tolylene diisocyanate (TDI), 4,4′-diphenylmethane diisocyanate (MDI), and 1,6-hexamethylene diisocyanate (HDI) which are all known to be toxic having a severe effect upon the respiratory system. The maximum allowable concentration (MAC) level for TDI has been suggested as 0·01 parts per million. It is not therefore possible to form blocked adducts directly with these compounds since the unblocking at high temperatures would evolve unacceptable quantities of toxic vapours. However, these compounds can be converted to an acceptable form of polyisocyanate by prereacting excess diisocyanate with a polyhydric alcohol to yield a relatively harmless compound of low vapour pressure.

Commonly employed for this purpose is trimethylol propane (TMP) possessing three primary hydroxyl groups. Two equivalents of diisocyanate should be reacted with one equivalent of TMP.

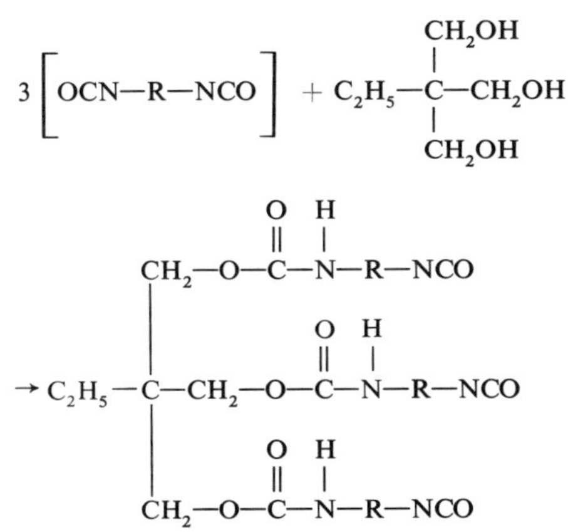

This isocyanate terminated prepolymer is then processed with a suitable mono hydroxy compound to yield the blocked isocyanate.

The choice of blocking agent is important for cross-linking powder coating resins since a volatile by-product must be evolved during the stoving operation. Phenolic materials have been used but these are unpleasant

Table 5. Some Commercially Available Curing Agents for Epoxy Powder Coatings

Type	Reference	Supplier	Amount parts/hundred/resin( )	Stoving schedule
Accelerated dicyandiamide	Epikure 107FF	Shell Chemical	5·0	180°C/15′
Accelerated dicyandiamide	Epikure 108FF	Shell Chemical	5·0	180°C/10′
Accelerated dicyandiamide	HT.2831	Ciba-Geigy	5·5	180°C/10′
Accelerated dicyandiamide	DEH 40	Dow-Chemical	5·0	180°C/10′
Accelerated dicyandiamide	DEH 41	Dow-Chemical	5·0	180°C/10′
Substituted dicyandiamide	HT.2833	CIBA-Geigy	3·8	180°C/15′
Accelerated substituted dicyandiamide	XB.2632	CIBA-Geigy	3·9	180°C/10′
Imidazole	B.31	Veba-Chemie	6·3	180°C/10′ or 140°C/15′
Polyamide	V.208	Veba-Chemie	33·7	180°C/15′
Polyamide	V.209	Veba-Chemie	28·0	180°C/15′
Acid anhydride	PMDA	Veba-Chemie	13·3	200°C/20′
Acid anhydride	XB.2622	CIBA-Geigy	12·4	180°C/10′
Accelerated acid anhydride	XB.2731	CIBA-Geigy	12·4	180°C/5′
Accelerated acid anhydride	LMB1521	CIBA-Geigy	12·4	180°C/3′
Anhydride adduct	XD.5566	Dow Chemical	18·0	180°C/10′

Film properties						
Gloss	Adhesion	Flexibility	Water Resist.	Chemical Resist.	Solvent Resist.	Colour
moderate	good	good	good	good	good	good
moderate	good	good	good	good	good	good
moderate	good	good	good	good	good	good
moderate	good	good	good	good	good	good
moderate	good	good	good	good	good	good
very good	good	very good	good	good	moderate	good
very good	good	very good	good	good	good	good
very good	good	good	very good	very good	very good	moderate
excellent	good	excellent	excellent	very good	poor	poor
excellent	good	excellent	excellent	very good	poor	poor
poor	good	moderate	good	good	very good	very good
poor	good	good	good	good	very good	very good
poor	good	good	good	good	excellent	very good
poor	good	good	good	good	excellent	good
moderate	good	good	very good	very good	very good	good

compounds to discharge to atmosphere, and they have been shown to cause blistering of the cured film.

PHENOLIC RESINS AS CURING AGENTS FOR EPOXY RESINS

A considerable investigation has been made of phenolic resins since they were first introduced as surface coating materials at the beginning of this century. Reaction between the phenolic nucleus and an aldehyde compound, most frequently formaldehyde, can take place as addition and condensation processes under the influence of both acidic and basic catalysts.

The initial stage of the reaction, shown below by reacting phenol and formaldehyde, is the formation, by addition, of the phenol alcohols. These may further polymerization by the condensation process of removing water and formaldehyde to yield highly cross-linked resinous structures.

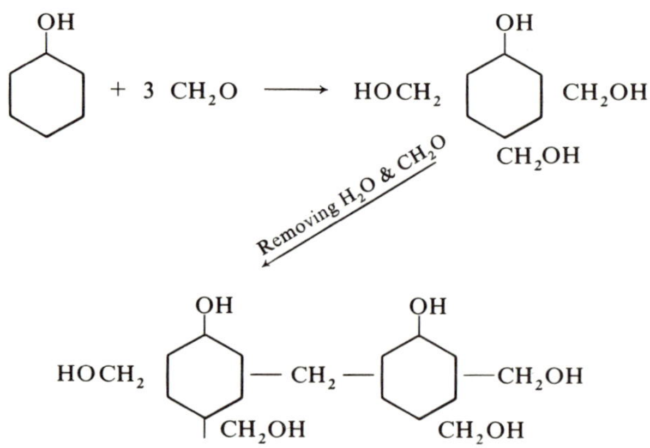

Ultimately yielding a cross-linked, three dimensional network.

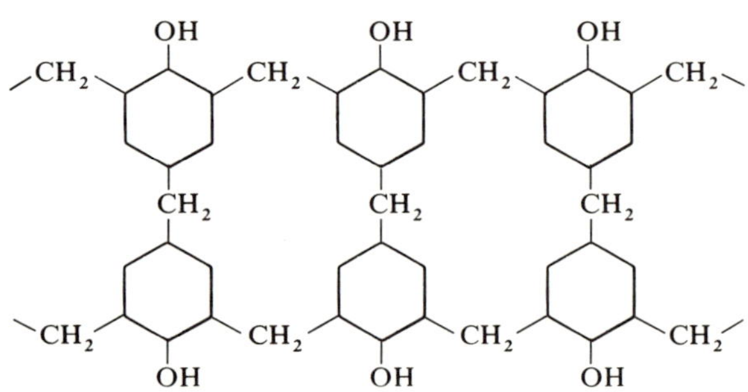

This is a greatly simplified picture of a series of involved chemical reactions.

Phenol itself has three reactive sites and can quickly form highly cross-linked resins. Substituted phenols are available with two reactive sites and the formation of linear polymer resins is possible.

The substituted phenols will react readily with formaldehyde at temperatures in the region of 60–80°C in the presence of a strong alkali to form a linear resinous compound which separates readily from the aqueous carrier of the formaldehyde. This may be decanted, and after several washings to remove traces of alkali, it can be dried out to give a toffee-like clear resin of fixed melting range.

The melting point is dictated by the quantity and type of reactants, including catalyst, and the reaction conditions. The nature of the substituent in any position on the phenolic nucleus will have considerable effect on the properties of the resin.

They will effect compatibility, colour stability, internal plasticity and, of course, the rate of reaction during subsequent heat curing or cross-linking with other resins.

It has already been mentioned that alkali metal hydroxides will promote the self polymerization of epoxy resins and that these are members of the phenolic resin family. Combinations of phenolic resins and epoxy resins can therefore be produced by the etherification process illustrated in the section concerned with the effect of Lewis bases. It is also possible that the joint reaction can be assisted by catalytic amounts of Lewis bases.

The reason for developing blends of epoxy and phenolic resin is the outlet in the can and drum lining field, where a product with improved water resistance and resistance to dilute mineral acids is required.

Cross-linking of epoxy resins with phenolic resins can occur by reaction of the phenolic hydroxyl or the methoxy groups still remaining in the phenolic resin.

AMINO RESINS AS CROSS-LINKING AGENTS FOR POWDER COATING RESINS

Amino resins suitable for curing powder coating resins may be produced by the reaction of formaldehyde with urea, melamine or benzoguanamine. Although the benzoguanamine resin may be discounted as a potential cross-linking agent for powder coating resins, because of its low reactivity.

Melamine resins have already been manufactured in a physical form suitable for powder coating compounding and they are preferred to urea resins because they impart better film hardness with improved alkali and detergent resistance qualities.

The reaction between melamine and formaldehyde takes place under slightly alkaline conditions to give methylol derivatives as the first stage. These then behave in a similar fashion to the phenol alcohols and a resinous compound results from a self-polymerizing condensation between methylol groups by removal of water and formaldehyde.

Theoretically one mole of melamine can add six moles of formaldehyde to form the hexamethylol derivative but in actual practice it is doubtful whether more than four moles of formaldehyde actually react.

D

Cross-linking between the melamine resin and epoxies, polyester and acrylics will take place with the removal of water to form an ether linkage with the hydroxyl groups present. Self-condensation in the melamine resin will also result in the liberation of traces of formaldehyde.

At the present stage of development melamine resins are not ideal components of powder coating systems. However, they are well known for their excellent qualities in certain conventional paint finishes and, no doubt, they will afford some technical properties not found in other powder coating systems.

The reaction between an epoxy and a melamine resin is shown in

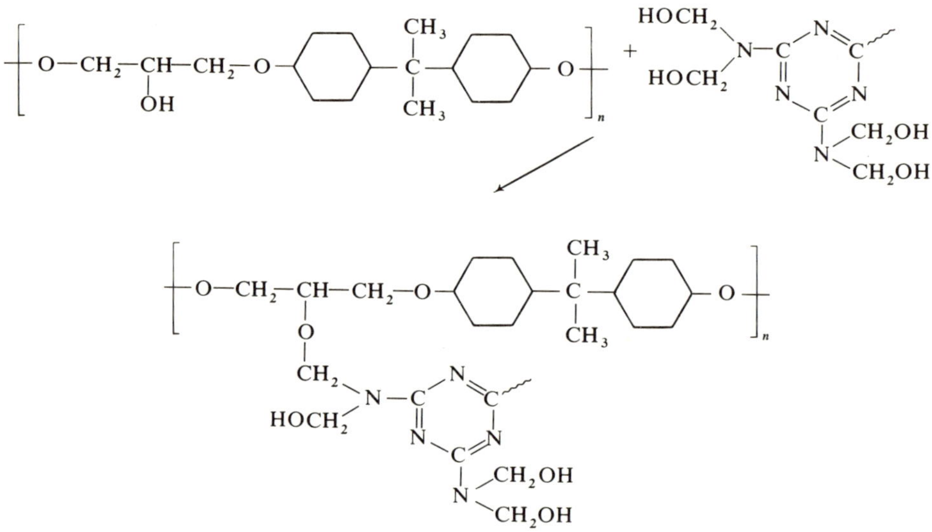

Similar reactions will occur with hydroxyl containing polyesters and acrylics.

HEXAMETHOXYMETHYLMELAMINE AS A CURING AGENT FOR POWDER COATING RESINS

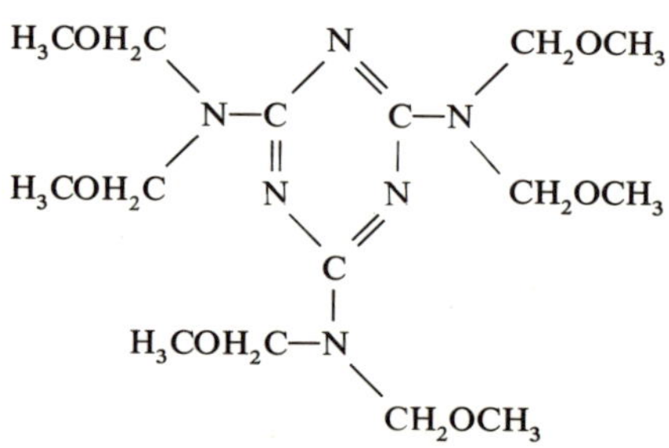

Hexamethoxymethylmelamine (HMMM), see formula above, reacts in acid conditions with hydroxyl and carboxyl groups as follows:

$$>N{-}CH_2\,OCH_3 + HO{\sim} \rightarrow\ >N{-}CH_2\,O{\sim} + CH_3\,OH$$

$$>N{-}CH_2\,OCH_3 + HOOC{\sim} \rightarrow\ >N.CH_2OCO{\sim} + CH_3OH$$

HMMM may be incorporated into a suitable polyester resin by melting both components in a Z-blade mixer at 80–90°C, the polyester being melted first and the HMMM added afterwards. After five minutes mixing, the mixture is removed, cooled and granulated.

EPOXY CROSS-LINKING AGENTS

The majority of the reactions discussed in this chapter have related to the use of epoxy resin as the main film forming resin, but it is often used in powder coating formulation as a cross-linking agent for polyesters and acrylic resins. For example:

(a) A commercial polyester resin with a high proportion of free carboxyl groups may be compounded with equal parts of an epoxy resin of equivalent weight in the range 700–900. It is claimed that this combination enhances the light and heat stability and the exterior durability of epoxies, without detracting from their adhesion qualities.[3]

(b) Polyester resins may also be cross-linked by up to 10% of a special epoxy resin—triglycidyl isocyanurate. Good gloss and excellent resistance to hydrolysis are claimed for this system. It does suffer, however, from a reduction in flow properties resulting in a pronounced orange peel effect.[4]

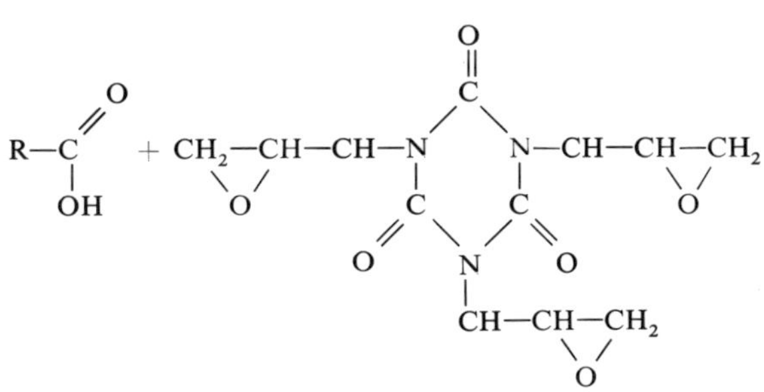

(c) Epoxy resins may be employed to crosslink acrylic resins containing carboxyl groups, viz:

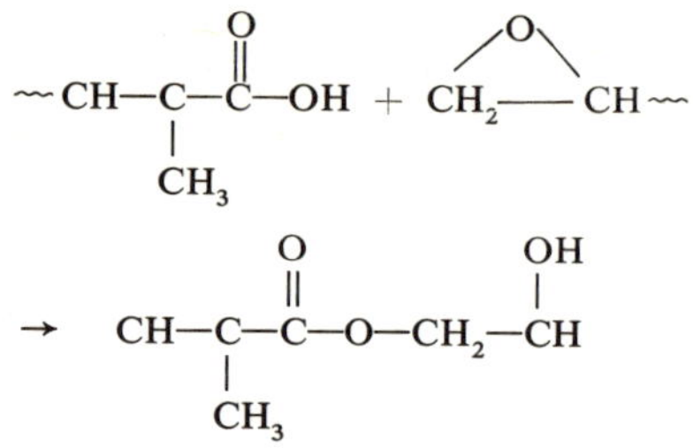

Generally it is preferred to introduce the glycidyl or epoxide group into the acrylic chain and effect the cross-linking by acidic resinous compounds.

PYROMELLITIC DIANHYDRIDE AS A CURING AGENT[5]

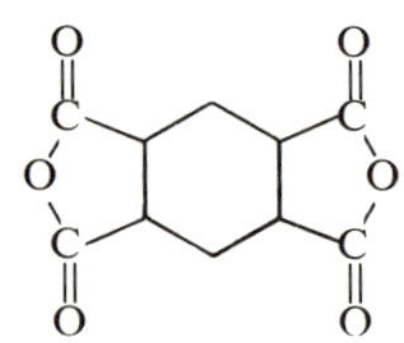

Pyromellitic dianhydride (see formula above) is an efficient curing agent for epoxy resins and some polyesters. Resins cross-linked by this agent, have improved heat stability and good chemical resistance.

The major disadvantage is its high melting point of 285°C, resulting in cured films with low gloss and poor flow, both due to the insolubility of the anhydride curing agent.

REFERENCES

1. Bayer. British Patent 1265634.
2. Textron. British Patent 1286422.
3. Scado.
4. Ciba-Geigy.
5. Veba Chemie. Technical Data Sheet.

4

The Formulation of Powder Coatings

SUMMARY

The formulator will generally use the basic formulating principles employed in the paint industry for tailor making a product to suit the end-use application. He will pay attention to the following essentials:—

1. The method by which the product will be compounded (see Chapter 6). A batch process in a Z-blade mixer can permit higher pigment loadings if desired and can obviate the necessity to masterbatch small quantities of liquid flow additives. The hardening agent used in the batch process must not react significantly during the extended period at elevated temperatures in the mixer.
2. Correct choice of pigment to maintain good rheological properties during the important premixing operation (see Chapter 5).
3. Correct choice of resin/hardener to give the required blend of film properties. It is important to take account of the toxic nature of the hardener, and the fineness of particle size needed to ensure good homogeneity in the compounding operations.
4. The need to ensure the correct colour match and the required level of gloss.
5. The correct setting of the fine grinding machine to produce the most suitable particle size distribution for the end-use application method (see Chapter 8).
6. A further important consideration is the possibility that new material might contaminate products already being manufactured.
7. Good storage stability and freedom from "caking" must also be considered.
8. Re-cycled powder must be capable of being applied without the need for reprocessing by the manufacturer, or mixing with excessively large amounts of new powder.

In fact he must generally formulate for low toxicity at the dusty premixing stage, for a product which will not react exothermically and cause gelation during the compounding process, and for a product which will not produce any risk of explosion during the fine grinding operation or during the subsequent application process at the customers plant.

Choice of Resin/Hardener and Flow Control Agent

The properties of the cured film will be largely influenced by the resin/hardener combination and this choice could be influenced by the equipment available for compounding the powder coating. Unless otherwise stated, it will be assumed that an extruder will be employed for compounding. The use of flow additives is included in this section because they influence the binder more than they effect the pigment.

GLOSS

This important property will be influenced by several factors.

The Choice of Resin

A low melting point resin will begin flowing at a lower oven temperature to give a smooth film before the curing agent becomes effective. It is logical therefore that there should be a good differential between the fluxing temperature and the temperature at which cure begins if the gloss is not to be impaired by poor flow. Excessive use of low melting point resins can cause blocking in the feed compartment of the extruder, so that these resins are best used in conjunction with the higher melting point resins. Thermoplastic resins produced by the polymerization of various hydrocarbons may be used in small quantities to enhance flow and gloss, but they must be employed cautiously to avoid deterioration of the mechanical and chemical resistance of the cured film.

The Choice of Hardener

For best gloss and flow the hardener must be compatible with the resin. Most of the accelerated dicyandiamide curing agents are not entirely compatible and the substituted dicyandiamides or imidazole curing agents are preferred. Acid anhydrides generally impart reduced gloss due to their high melting points and poorer compatibility. However, it is possible in some instances to improve the gloss of anhydride cured systems by extruding at slightly higher temperatures.

Flow Control Additives

These are usually resinous compounds which are slightly incompatible with the powder coating resins and therefore serve to correct any tendency to cissing or cratering defects caused by the surface tension of the film during fluxing. The optimum amount of flow control additive normally employed is 0·5–1·0% of the total resin. Quantities in excess of these limits can adversely affect the gloss due to the inherent incompatibility of most flow agents.

Compounding Procedure

A profound influence upon the gloss level is the efficiency of the extrusion process. The more homogeneous the dispersion the better is the gloss level.

Poor dispersion of intrinsically glossy formulations can produce semi-gloss finishes and, in fact, this method is often used to reduce gloss.

Reduced Gloss

Low gloss is best achieved in powder coatings by a specific choice of resin and hardener. It is virtually impossible to produce a matt finish by the use of extenders, for the high temperature of the compounding operation wets out the extender more efficiently than in normal paint milling operations. Increasing the amount of extender will only result in poor film flow without marked decrease in gloss. Hardeners which produce matt finishes are essentially substances which are incompatible with the resin and this quality is reflected in the poorer mechanical properties of matt finishes. Polyethylene waxes may also be used to reduce gloss levels.

EDGE COVERAGE

The Choice of Resin

Certain manufacturers of epoxies include in their range of powder coating resins, products which possess a higher than normal melting point than the standard grades, but with similar epoxide equivalents. The use of 10–15% of these resins will improve edge coverage without appreciably detracting from the flow properties. A typical resin of this class is the Dow experimental product XD 3542. An alternative resin to improve edge coverage is a polyvinyl butyral of relatively high viscosity (i.e. Mowital B60*).

The Choice of Hardener

A reactive hardener of average compatibility will give better edge coverage than the more compatible hardeners by merely reducing the flow characteristics of the resin. Accelerated dicyandiamide systems will therefore give better edge coverage than the substituted dicyandiamides.

Flow Control Additives

These will generally detract from good edge coverage, so that the recommended levels should not be increased.

Compounding Procedure

An efficient extrusion procedure will usually detract from good edge coverage unless the more efficient dispersion has caused thixotropy to develop in the pigments.

* Trade name of Hoechst Chemicals.

ADHESION

The Choice of Resin

Good adhesion is often directly attributable to the presence of a hydroxyl group in a powder coating resin. It has a relatively small volume for a polar group and is not, therefore, subjected to the same degree of steric hindrance encountered by the much larger carboxyl group. Improved adhesion will be shown by those resins which are not highly branched and which can allow polar groups to make contact with the substratum during the fluxing process. The epoxide resins exhibit better adhesion properties than do polyesters or acrylic resins. A close molecular weight spread will also improve the adhesion of the coating.

The Choice of Hardener

As stated above, for maximum development of adhesion properties, the resin must be allowed to flow out and wet the coated surface. The polar groups, in particular the hydroxyl group, must be permitted to form a good adhesive bond with the surface before it is subjected to attack by the curing agent.

If all the polar groups are cross-linked before those in contact with the surface have formed a strong adhesive bond, then partial or complete detachment of the film from the substrate can result. This fault may take several weeks to occur after application, and it is a known problem in certain rapid cure systems. The curing agent must therefore be sufficiently latent below 150°C to permit surface adhesion, and it is better to use slightly less than the stoichiometric quantity, rather than use an excess.

Flow Control Additives

The flow agent can adversely affect adhesion if used in considerable quantity above the specified amount. Silicone based flow additives are more prone to affect adhesion than the commonly used Modaflow*.

Compounding Procedure

An improvement in the efficiency of dispersion will significantly improve adhesion. Poor dispersion and partial curing in the compounding process will reduce the ability of the resin to wet the substratum.

FLEXIBILITY

The Choice of Resin

Resins of higher molecular weight in the epoxy series are generally more flexible than the low molecular weight members of the series since they have

* Modaflow is a trade name of Monsanto Chemicals.

fewer reactive sites and are therefore less highly cross-linked during the curing process. Partial esters of epoxy resins are also much more flexible due to the internally plasticized contribution made by the dimeric acid ingredients. Polyesters and acrylic resins based upon acrylate esters are generally more flexible than epoxies.

The Choice of Hardener

Amine hardeners do not impart very good flexibility to cured films. Imidine hardeners are better than accelerated or substituted dicyandiamide. Polyamide curing agents are very good for promoting flexibility. Acid anhydride curing agents give better flexibility than most amine hardeners.

Flow Control Additives

These do not detract from the flexibility of the cured film.

HEAT AND LIGHT RESISTANCE

The Choice of Resin

Epoxy resins have relatively poor resistance to discoloration by heat and ultraviolet light. Polyesters exhibit good light and heat stability especially those based on aliphatic polyols and polybasic acids. Acrylic resins also demonstrate good light and heat resistance unless they contain high proportions of styrene.

The Choice of Hardener

Epoxy resins show some improvement in heat and light stability when cross-linked with acid anhydride curing agents. Amine cured systems are extremely poor. Polyesters cured with dianhydrides are better than those cured with epoxy or amino resins. The acrylic resin gives poor results when cured with epoxies, but other hardeners generally give good results. Aliphatic diisocyanates impart good light and heat stability.

Flow Control Additives

"Modaflow" gives poor light and heat stability while silicone fluids are generally good.

Compounding Procedure

Efficiency of dispersion does not affect this film property.

CHEMICAL RESISTANCE

The Choice of Resin

Highly branched epoxy resins and those of high molecular weight give excellent resistance to alkalis. Polyesters are mostly poor for chemical resistance due to the presence of the easily hydrolysed ester linkage. Acrylic resins are generally good for chemical resistance unless they contain high proportions of acrylic esters. Phenolic resins give good resistance to dilute mineral acids.

The Choice of Hardener

The most reactive of the amine hardeners give the best chemical resistance so that imidines are preferred to dicyandiamide curing agents. Acid anhydrides are generally good curing agents for imparting chemical resistance. Diisocyanates are the best curing agents in this respect.

Flow Control Additives

These do not markedly affect this property.

Compounding Procedure

The efficiency of dispersion has an appreciable influence on chemical resistance. A homogeneous compound reacts more smoothly and crosslinks more efficiently, thereby improving chemical resistance.

CORROSION RESISTANCE

This property seems to be related directly to good adhesion and the comments on page 46 in this section should be reconsidered when formulating powder coatings for corrosion resistance.

WATER RESISTANCE

The Choice of Resin

Epoxies are not the best resins for water resistance although this property is greatly improved by incorporating a phenolic resin. Polyesters are poor and again they are susceptible to hydrolysis of the ester linkage. But acrylics exhibit good resistance.

The Choice of Hardener

The water resistance of all three of the main powder coatings resins will be greatly improved by crosslinking with a diisocyanate. Polyurethanes formed from polyester resins are very good for water resistance. Imidine curing agents are preferred to the other amine catalysts and to the acid anhydrides. Polyamides impart good water resistance to the cured film.

Flow Control Additives

Silicone based flow agents slightly improve the water resistance of powder coating films.

Compounding Procedure

Improved homogeneity, due to efficient dispersion, enhances the water resistance of the cured film.

DETERGENT RESISTANCE

Choice of Resin

Epoxy resins are quite good for resistance to detergents and an improvement can be found using the partial ester. Polyesters can given good resistance if formulated as suggested in Chapter 2. Acrylic resins are especially good in this respect in particular those resins based on acrylamide.

The Choice of Hardener

Diisocyanates and imidines are good curing agents for detergent resistance systems. The acid anhydride adduct system is also claimed to give good results.

Flow Control Additives

These have little effect upon this film property.

Compounding Procedure

Good homogeneity due to efficient compounding will enhance the detergent resistance of cured powder coatings.

EXTERIOR DURABILITY

The Choice of Resin

Epoxies must not be used in coatings required to give good exterior durability due to their pronounced tendency to chalking on outside exposure. Polyester resins are very good while acrylic resins are better than epoxies and in some cases as good as polyesters.

The Choice of Hardener

Diisocyanate cured and acid anhydride cured polyesters are very good for exterior durability.

Flow Control Additives

These do not greatly affect this property.

Compounding Procedure

The property is not appreciably affected one way or the other by more efficient dispersion.

TYPICAL STARTING POINT FORMULATIONS FOR POWDER COATINGS

The following formulations can be used as starting points from which successful formulations can be evolved. Final formulations must always be tested thoroughly in the laboratory before starting production.

High Gloss Decorative Epoxy—White

Araldite 7004[a]	61·5
10% Modaflow in 7004	3·2
Titanium dioxide	32·2
Hardener HT.2833[a]	3·1
	100·0

Instructions

The Modaflow is predispersed (see page 53) in resin and granulated to particles of 3–5 mm diameter. The resin is also preground to 3-5 mm granular size. All components are premixed (see Chapter 5) for ten minutes, then extruded from a Buss Kokneader PR 46 with a barrel temperature set at 75–85°C and a die temperature 95°C. The extrudate is cooled, granulated and fine ground (using a Mikropul Bantam to a particle size in the region of 30–75 μ. Stoving schedule: 180°C for 10 minutes.

Gloss White Epoxy with Good Edge Coverage

DER 663U[b]	43·0
5% Modaflow in 663U	10·0
XD.3542[b]	4·4
Titanium dioxide	40·0
DEH 41[b]	2·6
	100·0

[a] Araldite is trade name of Ciba-Geigy Ltd.
[b] Dow Chemical Co. Ltd.

Instructions

Granulate DER 663U, 5% Modaflow dispersion and XD.3542 to 3–5 mm diameter. Preblend all components for ten minutes, than extrude from a Buss Kokneader PR 46 with barrel temperature 80–90°C and die temperature 100°C. Cool the extrudate, granulate and fine grind to particle size 30–75 μ. Stoving schedule: 180°C for 15 minutes.

High Gloss White Epoxy, with Good Adhesion

Epikote 1055[(c)]	66·0
10% Modaflow in 1055	3·4
Titanium dioxide	27·5
Hardener Epikure 107FF[(c)]	3·1
	100·0

Instructions

Granulate 1055 and Modaflow dispersion to 3–5 mm diameter. Preblend all components for ten minutes. Extrude from the Eickhoff PWE 70 planetary extruder with barrel temperature 70°C. Cool the extrudate, granulate and fine grind to particle size 30–75 μ. Stoving schedule: 185°C for 15 minutes.

High Gloss White Epoxy with Good Flexibility

Araldite 7084[(a)]	40·0
10% Modaflow in 7084	5·0
Titanium dioxide	40·0
Veba-polyaminoamide V208[(f)]	15·0
	100·0

Instructions

Granulate 7084 and Modaflow dispersion to 3–5 mm diameter. Preblend all components for ten minutes. Extrude at a barrel temperature of 80–90°C and a die temperature of 100°C. Cool extrudate quickly, granulate and fine grind to particle size 30–75 μ. Stoving schedule: 180°C for 15 minutes.

White Polyester with Good Heat and Light Stability

Dynapol P.3000[(d)]	65·3
Acronal 700L[(e)]	2·1
Titanium dioxide	26·3
PMDA	6·3
	100·0

(c) Epikote and Epikure are trade names of Shell Chemical Company.
(d) Dynapol is a trade name of Dynamit-Nobel Ltd.
(e) Acronal is a trade name of B.A.S.F. Ltd.

Instructions

Preblend all components for ten minutes. Extrude at a barrel temperature of 100°C and die temperature 105°C. Cool extrudate, granulate and fine grind to particle size 30–75 μ. Stoving schedule: 190°C for 20 minutes.

White Epoxy with Good Chemical Resistance

DER 642U[b]	51·8
5% Modaflow in 642U	10·0
Titanium Dioxide	35·0
DEH 41[b]	3·2
	100·0

Instructions

Granulate 642U and Modaflow dispersion to 3–5 mm diameter. Preblend all components for ten minutes. Extrude at barrel temperature of 70–80°C with die temperature 90°C. Cool extrudate quickly, granulate and fine grind to 30–75 μ. Stoving schedule: 180°C for 10 minutes.

White Polyurethane with Good Water Resistance

Oxyester P.1652[f]	34·7
Veba Adduct B.1065[f]	34·7
Titanium dioxide	30·3
Silicone Oil OL[g]	0·3
	100·0

Instructions

Granulate P.1652 and B.1065 to 3–5 mm diameter. Preblend all components for ten minutes. Extrude at a barrel temperature of 100°C and die temperature 110°C. The preblending and extrusion may be replaced by compounding all ingredients in a heated Z-blade mixer at 110–120°C. Cool, granulate and fine grind to particle size 30–75 μ. Stoving schedule: 190°C for 15 minutes.

Gloss White Epoxy with Good Detergent Resistance

Shell 2057[c]	63·8
Titanium dioxide	30·0
Silicone fluid L.050[h]	0·2
Dimethyl terephthalate[i]	1·4
Veba B31[f]	4·6
	100·0

[b] Dow Chemical Company.
[f] Veba-Chemie. [g] Bayer.
[c] Shell Chemical Company. [h] Wacker Chemie. [i] I.C.I. Ltd.

Instructions

Granulate 2057 to 3–5 mm diameter. Preblend all components for fifteen minutes. Extrude through Buss PR.46 with barrel temperature 80–85°C. Screw temperature 25°C. Die temperature 90°C. Cool extrudate immediately, granulate and fine grind to 30–75 μ. Stoving schedule: 180°C for 10 minutes, or 140°C for 15 minutes. Store powder under cool conditions.

Gloss White Polyester with Good Exterior Durability

Scadonal S.2200[j]	60·0
XB.2615 GS[a]	6·7
Titanium dioxide	33·3
	100·0

Instructions

Preblend all components. Extrude at barrel temperature 80–90°C and die temperature 95°C. Cool extrudate, granulate and fine grind to 30–75 μ. Stoving schedule 180°C for 15 minutes.

FLOW CONTROL ADDITIVES

In the formulations outlined on previous pages a wide range of flow additives have been suggested. There is no single additive which is universally acceptable in all powder coating formulations and it will be necessary to assess each type and decide the optimum level of incorporation. Excessive use of flow additives should be avoided since the liquid materials can cause blocking of the stored powder due to "sweating out".

Typical flow additives include:

Modaflow liquid, produced by Monsanto. This is a highly viscous liquid which is best incorporated by a master batch technique. The resin is melted in a Z-blade mixer and the temperature kept below 120°C. Modaflow may be added as the viscous liquid but it is preferable to thin the product slightly with a small amount of toluene to assist the dispersion.

If the masterbatch is not cooled and emptied as soon as the Modaflow has dispersed, a deterioration in colour will soon be apparent due to the poor heat stability of Modaflow. This discoloration does not, however, seriously affect the performance of the Modaflow.

It is also possible to blend Modaflow directly at the high-speed preblending stage by metering the correct amount with a high pressure pump.

A solid form of Modaflow is also available and this is a fine powder which can be incorporated directly into the premix or even at the fine grinding stage to correct poor film flow of the final powder.

[j] Scado. [a] Ciba-Geigy.

Polyvinyl butyral of low viscosity such as the Mowital B.30H supplied by Hoechst Chemicals is widely used, at quantities up to 5%, in many epoxy powder coatings to improve flow and levelling.

Silicone fluids are less predictable in their levelling effects and are often found to be specific to one type of resin. Examples of these products are: Silicone Oil OL supplied by Bayer, Silicone fluid L.050 supplied by Wacker Chemie and Dow Corning products 470 and 471.

Flow control in polyester powder coatings may be imparted by the use of Cellulose Acetate Butyrate and this is described in a leaflet issued by Eastman-Kodak.[1] A further benefit gained from the use of CAB in polyester powder coatings is the improvement in free-flowing qualities and enhanced anti-caking properties.

There are several polymeric flow additives including Acronal 700L supplied by B.A.S.F. and the Borchigol products supplied by Borchers.

PIGMENTS FOR POWDER COATINGS

There are many pigments available to the paint manufacturer and they have not yet all been assessed for suitability in powder coatings. The technique of incorporating pigments into powder coatings, involving high temperature and high shear, differs considerably from the dispersion techniques available for wet paint systems, and the experience of the pigment manufacturers in the plastics field has to be reassessed in the light of the higher levels of pigment loading used in powder coatings.

In this section the pigments and extenders found to be suitable for use in powder coatings will be mentioned, together with comments relating to unsuitable pigments.

Titanium Dioxide

Many of the large manufacturers of titanium dioxide have assessed the suitability of the products in their range, in their own laboratories or in conjunction with the resin manufacturers.

In a series of reports prepared by the laboratories of Tioxide International[3] the following observations were made:

Differences were noticed at the processing stage of the powders and the pigments giving the best performance at the dry blending stage, generally gave the worst throughput rates during extrusion. A major problem of predispersion is the tendency for certain titanium dioxide grades to cake badly on the walls of the vessel so that the build-up has to be physically scraped off the walls from time to time during the premixing operation. If the titanium dioxide is heavily coated and possesses a large surface area then caking does not occur. The large surface area, however, makes it difficult to wet out the pigment in a solventless system, and high pigment loadings and the use of fast curing agents accentuates this defect.

A pigment loading greater than 0·5: 1·0 is difficult to extrude at an economic rate, although pigment loadings as high as 1·1 can be achieved in a Z-blade mixer.

There does not appear to be a great deal of difference in the performance of sulphate and chloride grades of titanium dioxide although the chloride grade is preferred because of its better colour.

The emulsion grades generally give low gloss in reactive systems and may be useful in the formulation of matt finishes.

The grades of titanium dioxide recommended for use in powder coatings include:

Tioxide TC.4	Tioxide International Ltd.
Kronos CL.220	Kronos Titanium Ltd.
Runa RO.676	Laporte Industries Ltd.

These are all chloride grades of titanium dioxide.

Lithopone

There is no difficulty with the incorporation of lithopone into powder coatings and it is found to improve the general film levelling properties. Coatings which are likely to be subjected to exterior weathering will tend to crack upon ageing when they contain a high proportion of lithopone. It is most suitable as a pigment for grey and pastel shades.

Carbon Blacks

There is little or no problem with the dispersion of carbon blacks. High carbon content channel blacks give good films at amounts of 2–3% of total resin. Finely divided blacks are problematic at the preblending stage and there is a strong case for preparing masterbatch dispersions of beaded carbons in resin only.

Tinting operations are best achieved with vegetable blacks but the development of colour is much more intense than the colour development in wet paint dispersion. For accuracy of addition as a tinting aid it is suggested that a dispersion of vegetable black is made as a 5% dry mix with barium sulphate.

Coloured Pigments

Several manufacturers of inorganic and organic pigments and dyestuffs have assessed the suitability of their pigments in epoxy powder coatings. These include I.C.I.,[3] Sandoz,[4] B.A.S.F.,[5] Reckitts.[6]

Ciba-Geigy are also able to advise on the suitability of certain pigments in their range.

Inorganic Colours

Yellow oxide and red oxide pigments are useful aids to tinting. Certain grades of yellow oxide show improved dispersion qualities in powder coatings.[6] Colour development of these oxide pigments appears to be more efficient in powder compounding processes and the amounts used for tinting are often considerably less than those used in obtaining a comparable wet paint shade.

Yellow and red chrome pigments are easy to incorporate in powder coatings[3] although the lemon chromes have poorer heat stability than the middle chromes and scarlet chromes.

Cadmium yellows and reds are easy to disperse in powder coatings and they exhibit good heat[6] resistance properties.

Organic Colours

Benzidine yellows impart poor flow properties to powder coating epoxy resins. Their performance with polyesters has not been assessed.

Toluidine reds are mostly unsuitable because they do not maintain their brightness in the cured film.

Small quantities can be used to adjust the shade of cadmium reds and scarlet chromes.

Phthalocyanine blues and greens are easy to disperse and they maintain their colour in the cured film.

Full development of colour in pastel shades may be achieved by predispersing white pigment and phthalocyanine by high speed continuous blending until the colour is constant. The resins, additives and hardeners may then be added.

Metallic Pigments

When aluminium or bronze pigments are compounded by extrusion into powder coatings they develop non-leafing qualities and it is difficult to produce a bright silver or gold by this method. Dry blending of metallic pigments into a preground base containing extender will give bright metallic finishes but it is not considered a safe practice to dry blend aluminium powders.

Extenders

It is possible to incorporate relatively large quantities of extenders into powder coatings without appreciable reduction in gloss. When the point is reached where the gloss is reduced it is often due to poor flow of the extended film. There is no strong preference for any class of extenders although barium sulphate has been found to be relatively trouble-free in the compounding operation.

TEXTURED FINISHES

The customers requirements for this type of powder coating finish may range from the fine wrinkle quality to the smooth organosol type of texture. Generally it is easier to produce the rougher wrinkle type finish by merely dry blending 10–20% of thermoplastic materials such as polypropylene. The grade and quality must be carefully chosen, for the high temperatures used to cure powder coatings can discolour many of the common thermoplastic resins.

Extenders can produce a slight texturing effect if they impart thixotropy to the powder coating. These are not usually well pronounced textures and resemble severe orange peeling.

HAMMER FINISHES

These powder coating finishes may be produced without the incorporation of silicone based hammer additives. Incompatible resinous compounds of low melting point can impart this effect and the formulator is advised to apply the general principles for formulating silicone free conventional hammer finishes.

ANTI-CAKING ADDITIVES

Mention has already been made of the use of cellulose acetate butyrate for improving the flow properties of polyester powder particles.

Small amounts of silicaceous extenders may be incorporated in the compounding operation or, better still, as a dry blended additive (0·05–0·1%) at the fine grinding stage. These silicas or silica gels act as an external lubricant and improve flow characteristics and prevent hard caking during storage.

REFERENCES

1. Eastman-Kodak. Formulator's Notes No. 3.16.
2. Tioxide International Ltd. Reports D.8635GC, D.8643 and D.8644.
3. I.C.I. Dyestuffs Division. Technical Information S.125.
4. Sandoz. Recommended shades to BS 2660 and DIN shades.
5. B.A.S.F.
6. Reckitts (Colours) Ltd. Data on yellow oxides, cadmium yellows and reds, ultramarine in powder coatings.

Manufacturing Processes

The following chapters deal with the different processes for powder coating manufacture.

Chapter 5—Premixing
6—Compounding
7—Granulation and Fine Grinding
8—Sieving and Classifying

SUMMARY

The success of spraying and curing a powder coating is measured by the appearance and performance of the finished coating.

It has been found that the manufacture of successful powder coatings depends upon a manufacturing sequence which is outlined in Table 6 and is detailed in the following chapters.

The aim is to achieve a homogeneous compounding of all the ingredients so that the final product is a powder coating of uniform particle size with each particle having the same composition.

It is difficult to incorporate and evenly distribute small quantities of essential additives or strong colours without predispersing these ingredients in the base resin by masterbatch techniques. This is simply an additional stage of hot melt compounding where the additives are melt mixed at known concentrations into a portion of the base resin. After cooling to solidify the masterbatch, this is granulated and it may then be added in the calculated amounts to the premix stage.

Table 6. Schematic Illustration of Powder Coating Production Procedures

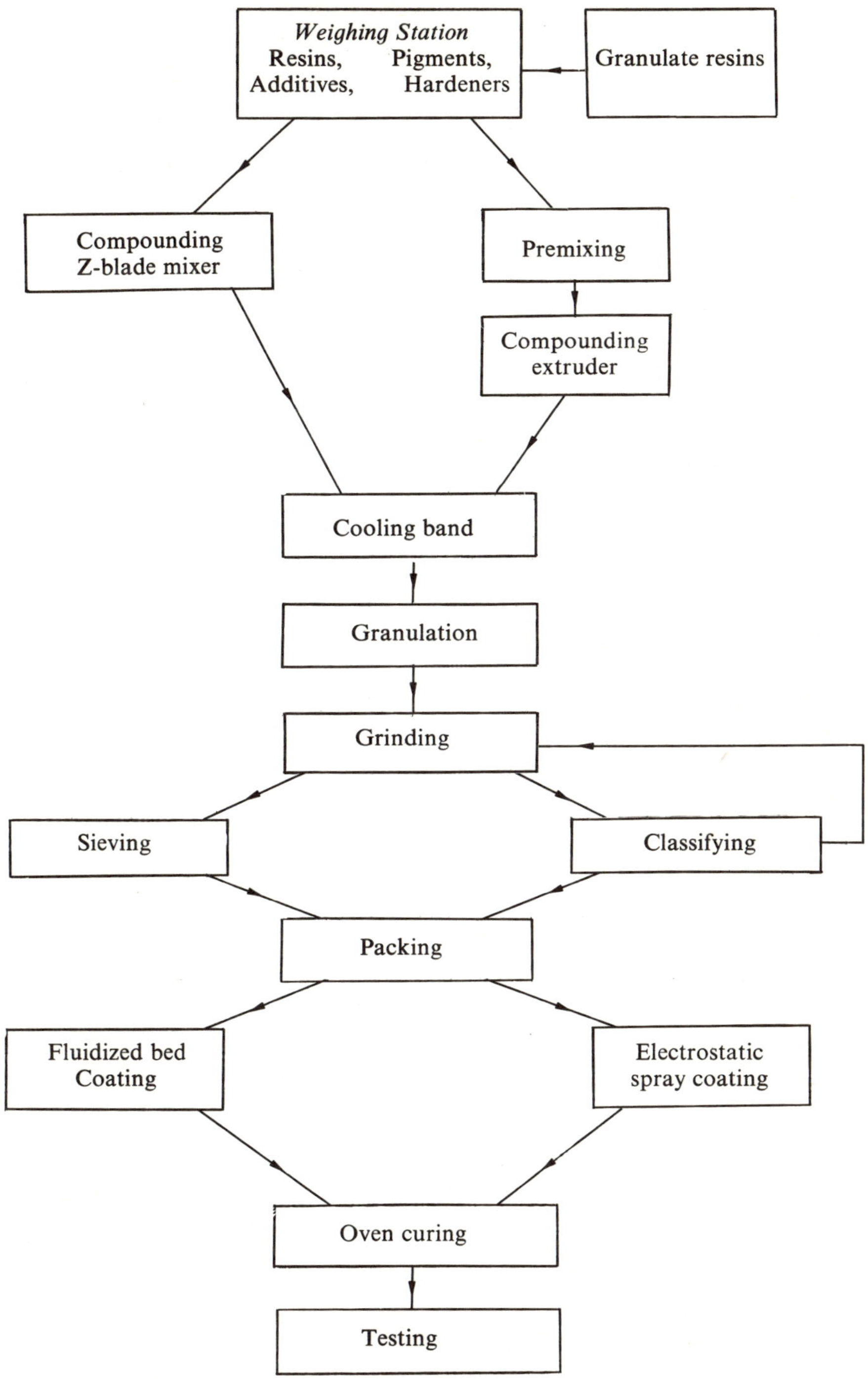

5

Premixing

SUMMARY

The first stage of the extrusion compounding operation is efficient preblending of components to ensure good homogeneity. Continuous extrusion processes, with dwell times in the extruder barrel of between thirty seconds and four minutes, must be served by a constant and uniform feed of well blended stock so that the product at the beginning of extrusion has the same composition throughout the entire production run.

In the early days, of manufacturing powder coatings, it was not generally realized that preblending operations were every bit as important as the extrusion stage. The present day emphasis on thin film coatings has underlined the importance of achieving homogeneity at all stages of manufacture, and significant improvements in film performance have been obtained simply by improving the efficiency of premixing.

Various techniques have been suggested for preblending and all the following processes are used for this purpose in production procedures in the powder coatings industry.

An important prerequisite to effective blending is uniformity in the particle sizes of the components. Homogeneity is best achieved with particles of comparable size, but in practice this is not realisable in view of the wide disparity in the particle sizes of resins and pigments as supplied. Unless masterbatch methods of compounding pigments, hardeners, and additives with resinous components are carried out prior to the preblending operation, the powder manufacturer will have to contend with the problem of attaining good homogeneity starting with a mixture of particles ranging from submicron size to pellets of ten millimetre diameter.

It must be appreciated that it is not enough to blend thoroughly in the premixing machine for the problem remains that the mixture will have to be transported to the feed hopper of the extruder and it is likely that some segregation of particles will occur during transportation and at the walls of the feed hopper.

This consideration makes a strong case for masterbatching of fine components; the advantages of handling a product of uniform particle size will be shown by:

(a) Good uniformity of premix.
(b) Ease of transportation and feed to the extruder.
(c) Rates of extrusion will be considerably improved.
(d) Cleaning of the mixing machine will be simplified.

Masterbatch methods, however, are costly and they are not essential for the successful manufacture of most standard powder coatings.

PREMIXING MACHINES

The following types of mixer are available:

Tumbler mixers.
Horizontal mixers.
Conical mixers.
High speed blenders.
Air jet mixers.

The advantages and disadvantages of each type will be considered.

Tumbler Mixers

DRUM HOOP MIXER

This device enables steel drums, in a wide range of sizes, to be used for the preblending operation. Open-ended drums are charged and sealed. A wide circular hoop is placed round the drum at an oblique angle then the device is mounted on two rollers at floor level. The desired mixing time is preset and fairly effective blending can usually be achieved in fifteen minutes.

Advantages

1. Low capital cost. Several of these devices can be purchased for the price of one mixing machine.
2. No dust is produced during the blending operation.
3. There is no delay between charging and mixing separate batches, i.e. several drums may be used and prepared in readiness for mixing.
4. No delays due to changing colour.
5. Easy to clean drums, by vacuum cleaner or water washing.
6. Can be stored away when not in use.
7. Operates on single phase wiring.
8. Drums up to 200 litre capacity may be easily handled by one operator.
9. The drums may be used for storing the premix or for transporting the premix to the feed hopper of the extruder.
10. It occupies very little floor space during operation and is light and easy to handle.

Disadvantages

1. Finely divided titanium dioxide and extenders adhere to the walls of the drum in layers several mm thick.
2. The effectiveness of dispersion is not good and small amounts of tinting pigments are not easily dispersed.
3. The rollers are subject to a high degree of wear and need to be replaced from time to time.
4. The drum must contain an optimum load to ensure good weight balance. Reduced charges may cause hoop to topple off the rollers.

Fig. 1. Combined drum and hoop (Courtesy J. Engelsmann A.G.)

Fig. 2. Base for drum hoop mixer (Courtesy J. Engelsmann A.G.)

Table 7. Range of Drum Hoop Mixers
(Courtesy J. Engelsmann A.G.)

1. Measurements	Standard unit	Special unit
(a) Base without motor and housing	$39\frac{3}{8}$ in × $19\frac{5}{8}$ in 1,000 × 500 mm	$41\frac{3}{8}$ in × $19\frac{5}{8}$ in 1,250 × 500 mm
(b) Hoop for up to 20 gallon (75 litres) drums	$31\frac{1}{2}$ in × $12\frac{1}{4}$ in 800 × 310 mm	$31\frac{1}{2}$ in × $12\frac{1}{4}$ in 800 × 310 mm
Hoop for between 20 and 50 gallon (190 litres) drums	$42\frac{1}{2}$ in × 17 in 1,080 × 430 mm	$42\frac{1}{2}$ in × 17 in 1,080 × 430 mm
(c) Foot switch including mounting flange	$4\frac{3}{8}$ in × 5 in 100 × 130 mm	$4\frac{3}{8}$ in × 5 in 110 × 130 mm
(d) Foot switches for flameproof motors	5 in × 7 in 130 × 180 mm	5 in × 7 in 130 × 180 mm
2. *Motors*	0·5 h.p. 280 r.p.m.	0·75 h.p. 280 r.p.m.

3. *Weights*	
Roller track, standard unit, without motor and housing	62 lb—28 kg
Roller track, heavy duty unit, without motor and housing	103 lb—47 kg
Motor 0·5 h.p. with housing	37 lb—17 kg
Motor 0·75 h.p. with housing	37 lb—17 kg
Motor 0·75 h.p. flameproof with housing	39 lb—18 kg
Drum hoop, standard unit $31\frac{1}{2}$ in (800 mm) diameter	33 lb—15 kg
Drum hoop, heavy duty unit $31\frac{1}{2}$ in (800 mm) diameter	37 lb—17 kg
Drum hoop, standard unit $42\frac{1}{2}$ in (1080 mm) diameter	52 lb—24 kg
Drum hoop, heavy duty unit $42\frac{1}{2}$ in (1080 mm) diameter	70 lb—32 kg
Drum Hoop, special $47\frac{1}{2}$ in (1200 mm) diameter	110 lb—50 kg
Foot switch and mounting, standard motors	$2\frac{1}{4}$ lb—1 kg
Foot switches (2) and mounting, flameproof motors	$4\frac{1}{2}$ lb—2 kg

4. *Drum Hoops*	*Speed*
Hoop $31\frac{1}{2}$ in (800 mm) diameter	32 r.p.m.
Hoop $42\frac{1}{2}$ in (1080 mm) diameter	26 r.p.m.
Hoop $47\frac{1}{4}$ in (1200 mm) diameter	22 r.p.m.

DRUM TUMBLING MACHINES

These devices are more permanent fixtures than the drum hoop mixer and can handle more than one drum if required. Most comments relating to the drum hoop mixer apply generally to the tumbling machine, they are normally wired three phase, they are heavy, not easily moved and are stable.

Table 8. Range of Drum Tumbling Machines
(Courtesy Winkworth Machinery Ltd.)

1 h.p. up to 100 lb (45 kilos) per drum
2 h.p. up to 180 lb (82 kilos) per drum
3 h.p. up to 300 lb (136 kilos) per drum
5 h.p. up to 400 lb (182 kilos) per drum

Fig. 3. Drum tumbling machines (Courtesy Winkworth Machinery Ltd.)

OBLIQUELY ANGLED CONICAL BLENDERS

These are produced for the tumble blending of powders or granular materials and the design may be varied to suit the customers requirements. The offset angle of the cone imparts a swirl or spiral movement and efficient blending is claimed without the use of internal blades or baffles. They may be jacketed for heating or cooling. A feature of this type of blender is the simple-to-operate mucron valve for discharging the mix when blending is complete.

Advantages

1. Fairly simple to operate and maintain.
2. Since they do not contain baffles they are not very difficult to clean.
3. They are not dusty in operation.
4. Medium price range.
5. May be heated or cooled as required.

Disadvantages

1. The efficiency of dispersion is not good when relatively small amounts of tinting pigments need to be incorporated.
2. A build-up of finely divided pigments occurs at the walls.

DOUBLE CONE MIXERS

This machine is similar in operation to the oblique cone blender. It rotates round a central axis and the mixing operation is promoted by baffles inserted at angles within the rotating drum. This type of blender is widely used for repetitive mixing in the plastics industry (see page 67).

Advantages

1. It is simple to instal and operate.
2. It may be jacketed for heating or cooling purposes.
3. Medium price range.

Fig. 4. Oblique cone blender illustrating Mucron valve (Courtesy Baker Perkins Chemical Machinery Div.)

Fig. 5. Oblique cone blender (Courtesy Baker Perkins Chemical Machinery Div.)

Table 9. Range of Oblique Cone Blenders
(Courtesy Baker Perkins Chemical Machinery Div.)

Work capacity litres	cu. ft.	Size diameter mm	ft.	in.	h.p.	Blender r.p.m.	Overall length mm	ft.	in.	Overall width mm	ft.	in.	Overall height mm	ft.	in.
29	1·0	406	1	4	½	40	990	3	3	815	2	8	1,500	4	11
71	2·5	534	1	9	¾	37	1,615	5	3½	990	3	3	1,245	4	1
142	5·0	685	2	3	1	32	1,800	5	11	1,295	4	3	1,395	4	7
283	10·0	838	2	9	3	29	1,880	6	2	1,600	5	3	1,600	5	3
707	25·0	1,181	3	10½	7½	24	2,465	8	1	2,285	7	6	2,285	7	6
1,415	50·0	1,448	4	9	15	21	2,740	9	0	2,640	8	8	2,640	8	8
2,122	75·0	1,675	5	6	20	19	3,225	10	7	3,220	10	6	3,220	10	6
2,830	100·0	1,905	6	3	30	18	3,660	12	0	3,660	12	0	4,010	13	2
4,245	150·0	2,136	7	0	50	17	3,960	13	0	4,060	13	4	4,060	13	4

Speeds and powers given above are based on materials weighing 75 lb/cu. ft.

Disadvantages
1. The efficiency of dispersion is not good when small amounts of tinting pigments need to be incorporated.
2. A build-up of finely divided pigments can occur at the walls and on the baffles.
3. It is difficult to clean, especially the baffles, and is best employed to blend a succession of batches of the same formulation.

Fig. 6. Double cone blender (Courtesy Gardners of Gloucester Ltd.)

Horizontal Mixing Machines

There are many different types of horizontal mixer but common features are body (ranging from a semi- to a full cylinder) and a horizontal rotating shaft carrying the mixing blades. The configuration of the blades is the major difference between different makes. For example, the Lödige mixer has a sequences of scoop ended blades within a completely cylindrical body. Efficiency of mixing within this machine may be assisted by baffles. The machine is usually top loading and bottom emptying.

The most common design of mixing device is the ribbon blade of single or double helical-spring configuration. This spiral design may be continuous or interrupted.

A special contra-flow blade (as illustrated in fig. 10) is claimed to give efficient mixing since the inner blade impels material outwards from the central shaft while the outer blade impels material inwards from the outside.

The interrupted blade impels the mixture at the top of the cylinder in one direction along the shaft and draws the mixture at the bottom in the opposite direction. This action, together with the tangential flow induced, ensures good and speedy dispersion.

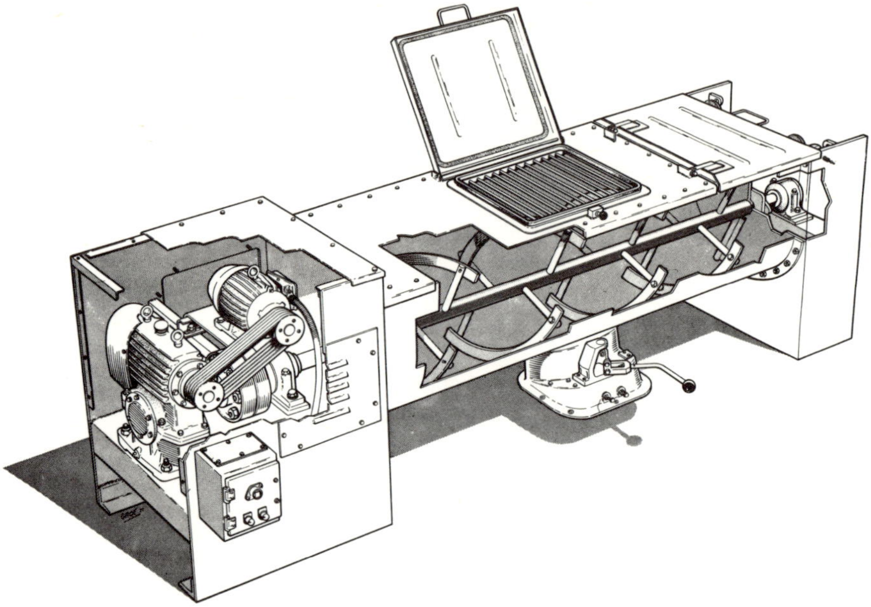

Fig. 7. Diagrammatic view of horizontal mixer (Courtesy Gardners of Gloucester Ltd.)

Fig. 8. Horizontal mixer (Courtesy Gardners of Gloucester Ltd.)

Fig. 9. Contraflow blender (Courtesy Winkworth Machinery Ltd.)

Advantages

1. Simple to instal and operate.
2. May be easily opened for cleaning and the shaft can be removed to allow good cleaning access to cylinder and emptying valve.
3. Medium price range.
4. Dispersion is adequate for most powder blending purposes including the incorporation of small amounts of pigments.

Disadvantages

1. They are rather more dusty in operation especially when the lid and end seals begin to wear.
2. Build-up of finely divided pigment occurs on the blades and at the top of the cylinder.

Conical Mixers

This type of blender comprises a single or multicone static assembly with a helical blade which rotates on a shaft which itself moves round the wall of the container. A three dimensional mixing action is obtained by the following mechanical action.

(a) Due to the movement of the screw about its own axis the materials are slightly mixed as they are transported upwards alongside the wall of the container.

(b) As the screw moves round the wall of the container the materials are transported simultaneously throughout the whole height of the container in circles, so that every particle participates in a conical spiral movement.

(c) The weight of the material opposite the mixing screw causes a downwards flow so that it blends together at the bottom of the conical container. As the material to be mixed remains closely together in one mass, no separation is caused by differences in particle size or bulk density.

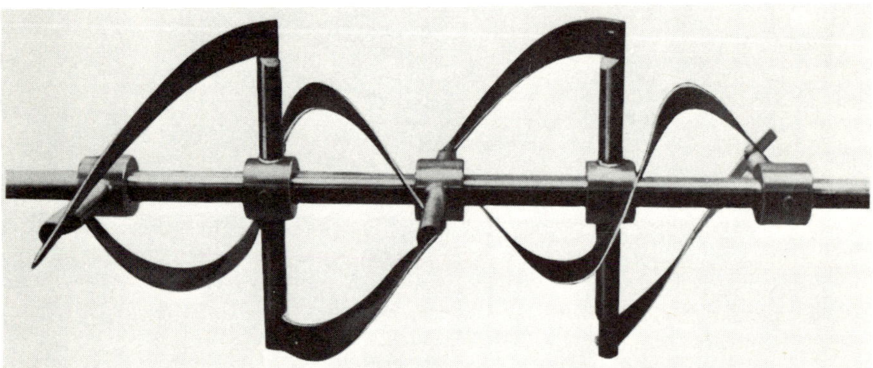

Fig. 10. Special Contraflow blade (Courtesy Winkworth Machinery Ltd.)

Advantages

1. It is a fast method for mixing large quantities.
2. It is a large capacity machine which can also handle relatively small mixes if required.
3. Efficiency of mixing is good.
4. Relatively low power consumption.
5. Liquids can be easily incorporated.
6. It is easy to clean.
7. Mixtures possessing poor flow are quite easy to discharge.
8. There is little heat development.
9. Medium price range.

Disadvantages

1. A large conical mixer is quite tall and for best use a two storey building or staging is necessary.

Fig. 11. Conical Mixer (Courtesy Nautamix N.V.)

An even more efficient version of the conical mixer is the combined cone mixer which has two or more cones joined to form one container with screws operating in each conical section.

Additional mixing occurs with the continuous transfer of mix from one cone to another.

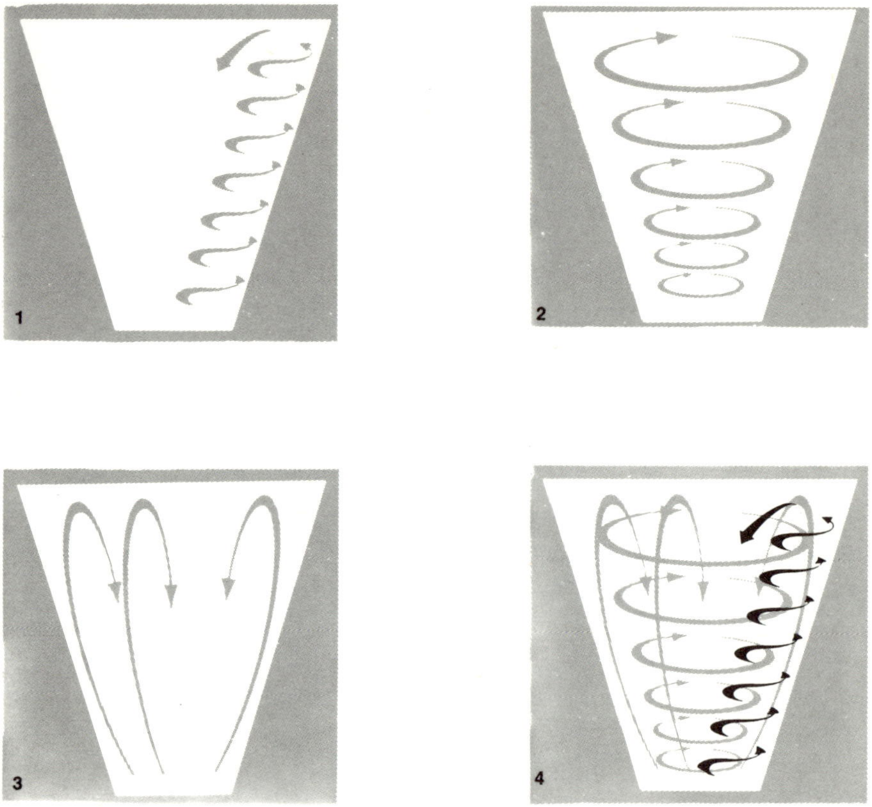

Fig. 12. Mixing operation of conical blender (Courtesy Nautamix N.V.)

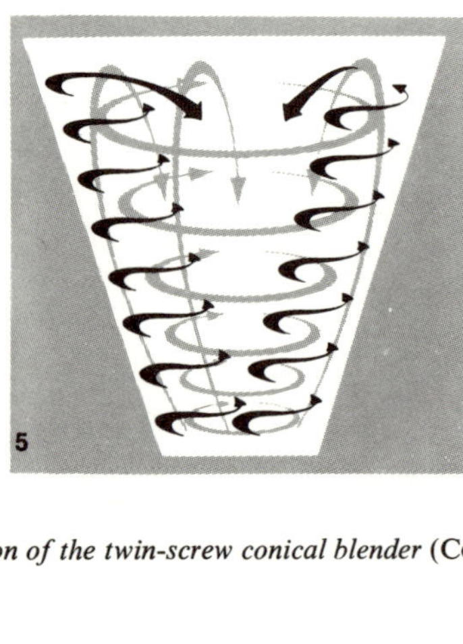

Fig. 13. Mixing operation of the twin-screw conical blender (Courtesy Nautamix N.V.)

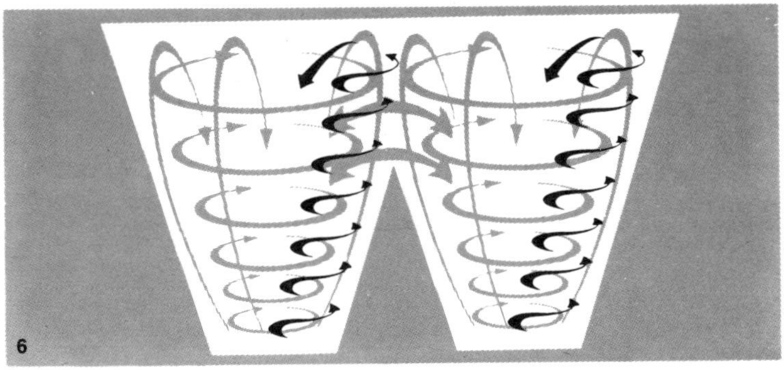

Fig. 14. Mixing operation of the combined cone blender (Courtesy Nautamix N.V.)

The High-Speed Blender

The principle of high-speed mixing is widely used in current European powder-preblending operations. A list of manufacturers is given in the suppliers section.

Basically the machine consists of a vertical cylindrical vessel at the bottom of which is a high-speed angled turbo-impeller.

The high-speed impeller causes the charge to be mechanically fluidized and flow round the vessel.

Dispersion is obtained by collision with the impeller, collision with the vessel wall, and attrition between particles as they circulate at high speed. A baffle may be incorporated to improve the already high degree of dispersion.

Advantages

1. High speed mixers will break down agglomerates of particles of fine powders which are merely dealt with as single particles by the slower mixers discussed so far.
2. The rapid mixing action will assist the full development of colour when tinting with relatively small amounts of dyestuffs. This can considerably reduce the quantities of, say phthalocyanine blues required to obtain a pastel shade.
3. Heat is evolved very rapidly and may be used to drive off water from titanium dioxide or extenders.
4. The equipment is very easy to clean.
5. Liquids are easily and homogeneously dispersed.
6. Mixing time is kept to a minimum.

Disadvantages

1. High power consumption.
2. High capital cost.

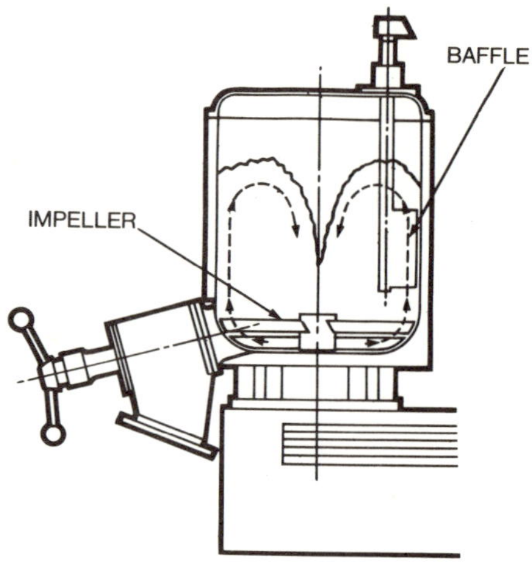

Fig. 15. Principle of the high-speed Drydisperser (Courtesy Baker Perkins Chemical Machinery Div.)

Fig. 16. The Drydisperser (Courtesy Baker Perkins Chemical Machinery Div.)

3. Wear on impeller is considerable.
4. May require cooling as additional expense.

Air Jet Mixers

Although these are not widely used in the powder coatings industry today they are included for completeness. While the other types of premixer may be employed with varying degrees of efficiency to mix granulates it is likely that the air jet mixer is more efficient for handling fine powders.

Advantages

1. Efficient blending of fine powders for tinting, anticaking, reclaiming of new and old powders.
2. There are no moving parts so wear is kept to a minimum.

Disadvantages

1. Usually occupy a large volume.
2. They are expensive.
3. They require large amounts of compressed air which is quite expensive.

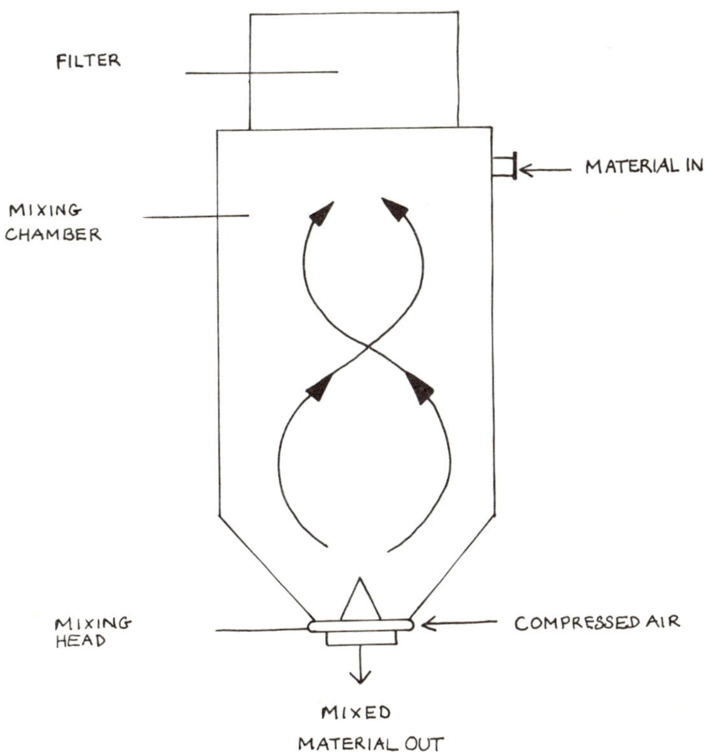

Fig. 17. The Airmix principle (Courtesy Gardners of Gloucester Ltd.)

DRY GRIND MANUFACTURE OF POWDER COATINGS

This method was developed some years ago in the United States for the manufacture of powder coatings with highly reactive curing agents, and the end products were mainly used as pipe coatings.

Since there is no hot melt compounding stage it is rightly included in the section relating to dry blending.

Pipe coating formulations call for good mechanical properties and resistance to "cathodic disbonding". Early manufacture was simply achieved by charging all the ingredients into a pebble mill, grinding for at least eight hours and discharging.

While the method at least had the advantage of simplicity, the disadvantages greatly outweighed this consideration for it was difficult to clean down to change colour. Furthermore the poor pigment dispersion resulted in coatings of inferior gloss and appearance.

Additionally the mill often required refrigerated cooling in summer and control of particle size was impossible.

In spite of considerable efforts to improve the process by altering grinding media, charge/mill-size ratio and the milling speeds and times, the powders still contained excessive proportions of fines which caused problems with spray application.

If today a requirement arose for dry grinding, a successful technique would probably be evolved with the aid of the high-speed mixer.

References to the all equipment described in Chapter 5 will be found in the section relating to Suppliers.

6

Hot Melt Compounding of Powder Coatings

SUMMARY

This manufacturing operation completes the homogenization of the powder coating ingredients which was begun in the preblending stage.

Resins used for powder coatings melt in the region of 75–105°C to give liquids which are highly viscous within this temperature range. Pigments, additives and hardeners are subjected to high shear forces during the period of the melt mixing. This results in the breakdown and dispersion of pigment agglomerates, the melting of the resin particles and the compatible distribution of flow additives and hardeners evenly throughout the resin media.

The efficiency of the shearing action, the working temperature and the time taken during compounding will be the major factors affecting the optimum development of film characteristics. However, the reactivity of the hardener system will largely dictate the temperature and time factors.

Compounding processes fall into three classes:

(a) The batch process, typified by Z-blade mixers.
(b) The semi-continuous process using heated rolls or Banbury mixers.
(c) The fully continuous manufacture by extrusion.

The large scale manufacture of top quality, thin film, powder coatings at high production rates, will dictate the purchase of continuous extruders and the correct choice of machine is of prime importance to manufacturers who are considering powder coating production. It is, far and away, the most costly item of manufacturing expenditure. It is also the operation which is most likely to dictate the rate of production.

COMPOUNDING TECHNIQUES

Batch Compounding

The sigma or Z-bladed mixer has been widely used for the heavy duty compounding of pastes since the beginning of this century and the equipment has proved to be simple to operate and sufficiently robust to require little servicing.

There are many variants of this type of machine since its operating principle has been adapted to suit a wide variety of different compounding processes. The type that is generally considered to be most suitable for the compounding of powder coatings (see fig 18) has a trough made up of two semi-cylindrical troughs welded together with the intersection forming a saddle piece. The blades sweep the whole of the working surface during each revolution, at the same time dividing the batch continuously across the saddle piece to achieve a perfectly homogeneous mix. Important design features required for powder coating production are:

(a) The trough must be jacketed for heating and cooling. Heating may be electric, by oil circulation or pressurized steam. Electric heating can cause localized hot spots (which must be avoided) and cooling is difficult to apply; oil heated systems can only be indirectly cooled which results in delayed cooling; therefore steam heating with water cooling seem to be the best combination because both processes employ the same jacket and the heat transfer properties are most favourable.

(b) Mild steel is suitable for construction in most cases although stainless steel is to be preferred to avoid corrosion.

(c) High shear must be obtained when the product is cold and very viscous. A belt drive will slip badly when the machine is producing the high shear required for the compounding of powder coatings so that a direct drive must be employed. To avoid the necessity for star-delta starting sequences the main shaft should be driven through a traction type fluid coupling. This torque-limiting device will obviate the affects of momentary overloads and will protect both machine and drive motors.

(d) Emptying the machine is best achieved by using a hydraulically operated mechanism to tilt the trough so that charging and emptying operations can be performed via the lid. An inching device should also be specified to control the rate of discharge and a means of reversing the blade movement will also facilitate the emptying procedure.

(e) Safety cut-out switches should be fitted to the lid system, preferably to a hinged safety grid between the lid and the trough so that the lid may be raised without arresting the drive operation. It is important to be able to raise the lid without stopping the machine so that various additions such as tinting, addition of flow additive or hardener addition may be made, and to allow any moisture contained in the pigments to escape to a ventilator.

(f) Cleaning of the machine is best achieved with the use of solvents and all glands must be designed with seals that are not affected by solvents.

Using a Z-blade mixer, a typical manufacturing procedure is described for the compounding operation of an epoxy-resin based powder coating:

Charge a proportion of the resin only and begin heating. The quantity of resin used will depend upon the pigment type and loading; it will probably lie between 40 and 60% of the total resin charge.

Fig. 18. Interior of Double Z-blade mixer (Courtesy Baker Perkins Chemical Machinery Div.)

Table 10. Range of Double Z-blade Mixers
(Courtesy Baker Perkins Chemical Machinery Div.)

Machine identification and recommended working capacity			Total capacity of trough			Power input (h.p.)		
Imp. gal.	litres	cu. ft.	Imp. gal.	litres	cu. ft.	Light series	Medium series	Heavy series
5½	25	0·88	8	35	1·24	3	5	10
8	35	1·24	11·2	50	1·80	4	7½	15
15½	70·4	2·48	22	100	3·52	7½	15	25
22	100	3·52	32	145	5·1	7½	15	30
31	140·9	4·95	44·7	200	7	10	20	40
44	200	7·05	62	285·4	9·9	10	20	50
62	280	9·9	88·5	400	14·2	15	25	50
88	400	14·1	127	575	20	15	30	60
125	570	19·8	177	805	28·4	20	40	80
176	800	28·4	254	1155	40	25	50	100

After several hours the resin will have melted sufficiently for the blades to be put into operation. The pigment can then be added in small amounts so that each portion is "wetted out" before a further addition is made. A dispersion aid may be included at this stage to assist pigment wetting. When all the pigment or extender has been charged the temperature is lowered to around 80°C and a period follows when the dough-like mass is subjected to high shear. Tinting to an approximate colour match is best carried out at this stage. When dispersion is considered to be adequate, the temperature is raised, the remaining resin is added and thoroughly mixed in. A homogeneous mixture of all the resin and pigment has now been made and the product is of medium viscosity at a temperature between 100–120°C. The flow additive can now be incorporated. Care must be exercised with Modaflow since it can discolour if kept at high temperature for a long period. The batch is now cooled and at the same time the hardener may be added and mixed in. (If the hardener is more reactive than dicyandiamide it is best to avoid addition in the Z-blade mixer and to incorporate it by dry dispersion at the grinding stage.) With all of the ingredients now thoroughly dispersed, the Z-blade mixer can be tilted and emptied. If the hardener is not included the mixture may be poured into trays and allowed to cool on storage racks at room temperature. When hardener has been added the product must be poured on to a cooling band and roller to effect rapid cooling.

The empty machine should be allowed to cool with the blades still operating in reverse and tilted so that the remaining mix adhering to the walls will flake off due to its brittle nature

For cleaning the Z-blade mixer in preparation for changing colour, the machine is allowed to run with a small amount of solvent, which may be heated, and a scouring agent such as sand, crepe paper or polyurethane foam is added. Rag may also be used for scouring.

Advantages of the Z-blade mixer

1. It is simple to instal and does not require constant skilled supervision.
2. It does not require a preblending operation.
3. Less costly to buy and instal than extruders.
4. Moisture adsorbed by the pigments is allowed to escape during compounding.
5. Larger amounts of pigments and extenders can be incorporated by this compounding method.
6. Liquid additives are easily introduced without masterbatching techniques.
7. Colour matching may be achieved by stepwise tinting.
8. Resins do not need to be granulated before charging.
9. When mixture temperature is attained, mechanical heat will maintain the temperature with little or no external heat supply.

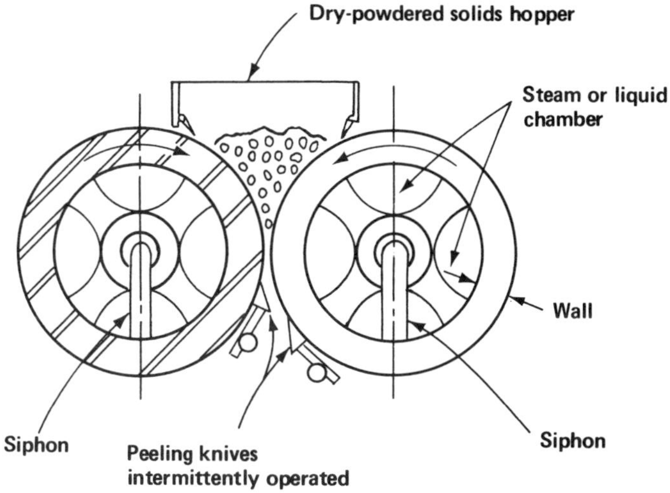

Fig. 19. The two roll mill

Disadvantages

1. Power consumption is high.
2. Dispersion of reactive hardeners is difficult to achieve without gelation.
3. The mixer is difficult to clean.
4. Large batches require lengthy processing times.
5. Emptying is a two man operation.
6. Production rates are lower than with extruders.
7. Gloss levels are lower than those of extruded powders.

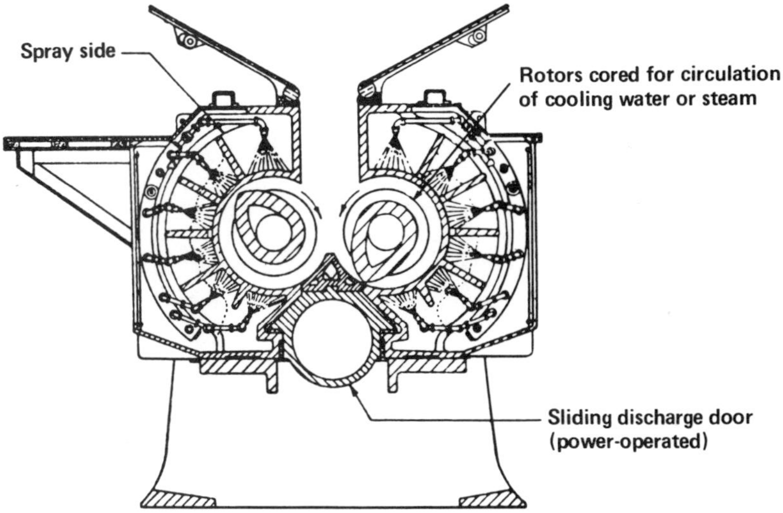

Fig. 20. The Banbury mixer

Semi-continuous Process

This method is not widely used for compounding powder coatings but is well known in the paint industry for the manufacture of coloured chip dispersions.

The use of heated two-roll mills or Banbury-mixers may be a feasible method for compounding thermoplastic powder coatings but it is not recommended for the manufacture of thermosetting powder coatings.

The Continuous Extrusion Process

The principles of extrusion compounding are best explained by direct reference to the main types employed in the powder coatings industry. Four different designs, two single and two twin screw extruders, have been found to give acceptable finished products; these are:

THE SINGLE SCREW EXTRUDER

The Buss Ko-Kneader is the most widely used single screw machine for compounding powder coatings. A characteristic feature of the Buss Ko-Kneader is the reciprocating motion superimposed on the rotation of the kneading screw which allows the specially profiled screw flights to knead the mixture axially against the three rows of stationary kneading teeth in the barrel casing.

The gear box illustrates the mechanical arrangement producing the double action of the kneading screw. This gear box is totally enclosed and possesses an automatic lubricating system which requires little maintenance.

The casing of the Ko-Kneader is usually manufactured in chrome stainless

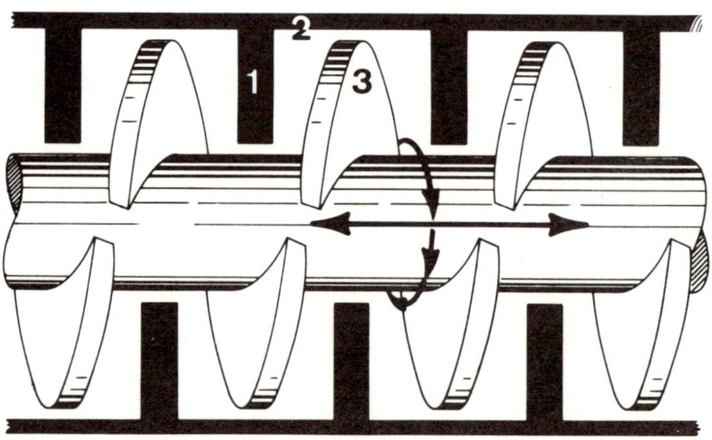

Fig. 21. Principle of the Ko-Kneader screw (Courtesy Buss Ltd.)

Fig. 22. Gear box Ko-Kneader (Courtesy Buss Ltd.)

Fig. 23. Open barrel and screw of Ko-Kneader (Courtesy Buss Ltd.)

steel with wear-resistant liners and it may be split vertically along the axis allowing the machine to be opened quickly to expose the whole interior for cleaning. Kneading bolts are fitted inside the casing and each one can be replaced should it become worn through excessive use.

Each kneading screw is machined from solid chrome stainless steel to fine tolerance limits; it is nitrided before the finishing operation in which both screw and casing are highly polished.

For heating and cooling; casings are provided with jackets which are subdivided into several zones for precise and individual temperature control. The screw is also heated but independently.

The inlet is fed from a hopper with a variable speed blade to prevent bridging in the premix. The level of the premix should never be allowed to fall below one fifth of the height of the hopper; should it do so a warning is sounded by depth monitors. For cleaning purposes the hopper is easily swung away from the inlet.

A specially designed die head is fitted to the exit end of the casing and this is jacketed to maintain close temperature control.

Separate heating units, with thermostats, are provided to control temperatures in the various zones.

Experience has shown that water heating and cooling is preferable to the indirectly heated circulated oil system and the original electric die heaters have now been replaced by water heated versions.

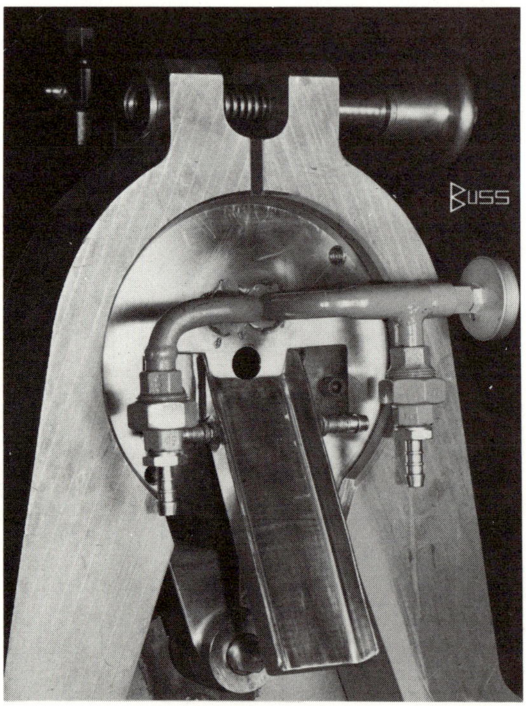

Fig. 24. Die head for Ko-Kneader (Courtesy Buss Ltd.)

The die head has been redesigned so that there is now little or no dead space behind the die head in which partially cured compound can collect.

Advantages of the Ko-Kneader

1. One trained man can operate several machines at the same time.
2. The power requirements are not high.
3. Good temperature control can be maintained and recorded.
4. It may be operated without the die head if required for very reactive systems.
5. Compounding efficiency is very good.
6. The self wiping action of the screw ensures that no dead spaces occur in the casing in which material can collect and polymerize.
7. The dwell time in the barrel is approximately two minutes.
8. It is very easy to clean and change colour.

Disadvantages

1. A finely divided premix feed drastically reduces throughput. It is not suitable for reprocessing finely divided powder.
2. If a metal object (a bolt) is accidentally fed into the machine, the kneading bolts may be sheared and the barrel damaged before the machine can be stopped. A safety overload cut-out would help to avoid excessive damage.
3. In common with most extruders, it needs to be fitted with a separate degassing extension if small amounts of volatile matter are given off during processing.

THE SINGLE SCREW EXTRUDER WITH PLANETARY ROLLERS

A range of extruders, manufactured by Eickhoff, embody a refinement of the single screw principle.

Table 11. Range of Buss Ko-Kneaders
(Courtesy Buss Ltd.)

Ko-Kneader types		PR 46	PR 100	PR 140	PR 200
Length	mm	7,500	9,500	10,000	14,500
Height	mm	2,100	2,400	2,850	3,200
Width	mm	3,000	3,100	3,600	4,000
Kneading screw diameter	mm	46	100	140	200
Max. driving power	kW	2	30	90	90
Screw heating/cooling zones		1	1	1	1
Casing heating/cooling zones		3	3	3	3
Heating capacity	kW	20	20	40	60
Adjustable output (approx. values)	kg/hr	5–30	150–400	350–800	750–1,500

Fig. 25. Ko-Kneader PR 140 with cooling/conveyor belt for the production of powder coatings (Courtesy Buss Ltd.)

The machine barrel is divided into two parts, one serving as a feed unit, the other as the extruder zone. The robust central shaft carries a forward-conveying screw in the feed compartment which transports the premix into the working zone of the extruder barrel. In this compartment both the cylindrical barrel and the central screw carry helical threads. The gap between the screw and the barrel contains six equidistantly-spaced fully-floating planetary rollers which again carry helical thread and slide easily into mesh with the grooves provided by the threaded screw and barrel. During the operation of the extruder, it is necessary to secure the die-head backing plate into position, to prevent any linear movement of the planetary screws, which would allow these to be ejected from the extruder.

The compounding of the premix is performed by the gentle wiping action of these planetary rollers.

Fig. 26. Schematic diagram of the planetary roller extruder (Courtesy Gebr. Eickhoff)

Fig. 27. PWE 70L with cooling rollers (Courtesy Gebr. Eickhoff)

The range of Eickhoff extruders runs from machines suitable for laboratory sampling to models designed for large scale production. The price of the laboratory model is comparable with that of the Buss Ko-Kneader PR 46 but has the advantage that its output ranges from a few kilos to fifty kilos per hour so that it can also be used for small scale production batches.

Heating and indirect cooling are applied to the barrel only of the laboratory model with an independent cooling system for the feed compartment. The die-head is electrically heated and thermostatically controlled. Larger production models may have independent heating for several zones of the barrel; the main screw may also be heated.

All machines are hopper fed with mechanical paddle agitation to prevent bridging of the premix.

Note: The earlier models of the Eickhoff range were often referred to as the Schalker extruders.

Table 12. Dimensions of Planetary Extruders in mm
(Courtesy Gebr. Eickoff)

	PWE 70	PWE 100	PWE 120	PWE 150
A	1,395–2,465	2,065–3,370	2,210–3,880	2,425–4,260
B	700	750	800	800
C	2,020–2,075	2,010–2,745	2,005–2,755	2,020–2,615
D	1,000	1,000	1,000	1,000

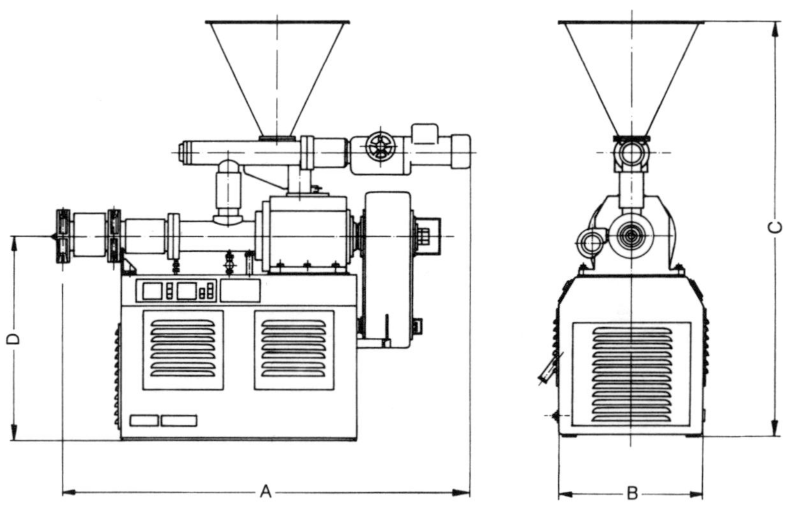

Fig. 28. Design features of the planetary extruders (Courtesy Gebr. Eickhoff)

Table 13. Technical Data of Planetary Extruders
(Courtesy Gebr. Eickoff)

		PWE 70	PWE 100	PWE 120	PWE 150
Drive capacity	kW	12/23	23/40	40/60	40/60/90
Max. screw speed	min^{-1}	50/100	30/50/80	30/50/80	35/50/80
Max. screw loading	J	2,300	7,350	12,750	12,750/17,250
Couple max.	(kpm)	(235)	(750)	(1,300)	(1,300/1,760)
Screw diameter	mm	70	90	120	150
LD ratio		6–19	5–23	6–22	5–21
Gearing length of roller section	mm	260	260	500	300
			500	700	500
			700		700
Number of planetary screws		6	6	8	10
Weight	N	11,770–16,680	17,650–31,400	25,500–41,200	25,500–45,750
	(kp)	(1,200–17,00)	(1,800–3,200)	(2,600–4,200)	(2,600–4,660)

Advantages of the Eickhoff Extruder

1. It is reasonably priced.
2. The efficiency of dispersion is quite good and suitable for most epoxy powder coating formulations.
3. It is extremely easy to change product colour because of the self cleaning action of the planetary rollers.
4. The barrel length is short and the dwell time in the extruder is not more than two minutes.
5. It does not require a lot of maintenance.
6. In most models the screw speed can easily be varied between 14 and 50 r.p.m. the highest speed is less for the larger models.
7. Power requirements are not high.
8. The space required for installation is not excessive.

Disadvantages

1. The barrel cannot be split for cleaning although the main screw and planet rollers are easily removed.
2. It is difficult to clean the feed compartment which tends to become choked due to premelting of the feed.
3. Indirect cooling does not provide the best temperature control.
4. The backing plate is difficult to extract.

THE TWIN SCREW EXTRUDER

The Baker Perkins multipurpose (MP) mixer is a most efficient machine for the continuous compounding of powder coatings.

A twin cylindrical barrel contains co-rotating screws with transverse lens-shaped sections set at ninety degrees, so that the screws at all times are wiping each other and the walls of the twin cylinders.

The diagram shows that three chambers of constantly changing volume are created by the movement of the lens shaped agitators. The material is

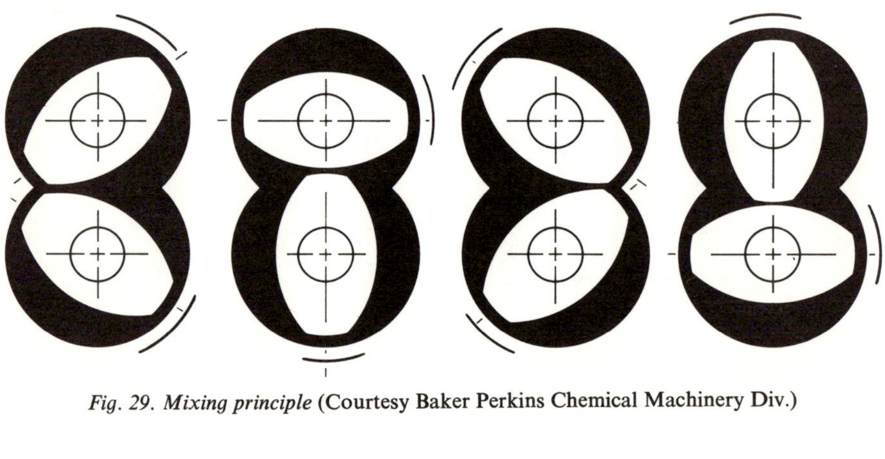

Fig. 29. Mixing principle (Courtesy Baker Perkins Chemical Machinery Div.)

therefore continually being wiped from the agitator screws and the barrel wall and transferred from one chamber to another. Excellent dispersion is achieved and the heat transfer is almost entirely adiabatic.

If the ninety degree relationship is maintained between the lens shaped cross sections it is possible to employ various screw profiles ranging from a helical continuous section of screw for blending and melting of the premix, to a set of discontinuous kneading paddles for the application of high shear to the melt-mix. These various profiles may be built in sections or machined from solid metal.

Each of the screws has provision for heating and cooling.

Fig. 30. MPC/V50 mixer (Courtesy Baker Perkins Chemical Machinery Div.)

The barrel has jackets over several zones for water heating and cooling.

An important feature of the machine is the axial movement of the barrel relative to the screw which can control the rate of extrusion at the discharge orifice thereby allowing an increase in back pressure and the intensity of compounding.

The discharge orifice may also be vented to allow gases to be bled from the mixer. This vent is installed at the low pressure end of the discharge section so that the product does not enter the vent.

For ease of cleaning, the barrel may be opened by a counter-balanced "clam shell" mechanism. In the barrel are several "blanked off" points which can be used to meter additions of liquid components during compounding.

The screw speeds are infinitely variable between zero and 400–500 r.p.m. depending upon the power rating of the drive-motor.

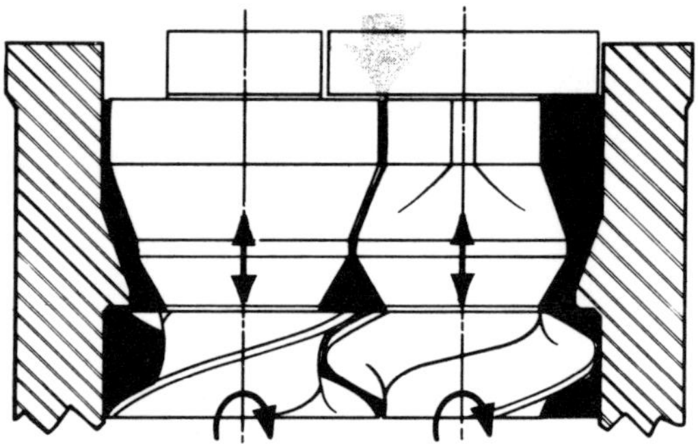

Fig. 31. Variable discharge orifice (Courtesy Baker Perkins Chemical Machinery Div.)

Premix is added by a vibratory feeder beginning with a slow rate of addition at low screw speeds. The feed rate is increased together with the screw speed and any tendency to overload is shown on a torque-meter. At the maximum throughput rate the screws will be running at top speed and a torque of between 80 and 100% will be shown on the dial. A sudden overload caused by the accidental addition of a "foreign body" will cut off power to the main drive and protect the machine from damage.

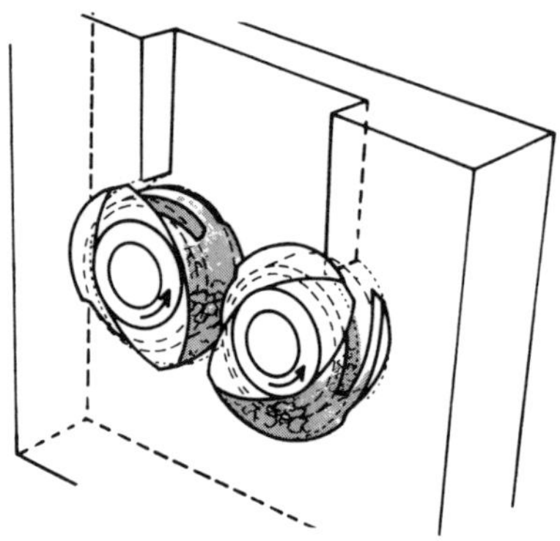

Fig. 32. Venting of discharge section (Courtesy Baker Perkins Chemical Machinery Div.)

The ability to relate feed rate to screw speed so that the maximum work load occurs in the compounding operation allows the optimum throughput to be achieved with premixed ingredients of widely different properties.

A typical start-up procedure for the MP mixer

(processing a standard epoxy powder coating)

An initial warm-up period of 60–90 minutes should be allowed before starting the run. Temperature controllers are set within the following limits:

Feed Hopper Jacket	Cold water
Zone 1 Jacket-adjacent to feed inlet	21–26°C
Zone 2 Jacket-adjacent to discharge	83–96°C
Discharge housing	83–96°C
Screws	60°C
Die Heater	83–96°C

The above temperature pattern may be applied to most epoxy based systems and the figures are carefully chosen to minimize the amount of heat transfer between the jackets and the processed material thereby maintaining an adiabatic process. Feed hopper jackets must be kept cold to prevent fusion of the premix in the non-compounding area. Zone 1 is also kept relatively cool to prevent premature heat fusion which would extend the residence time of the premixed ingredients and decrease the efficiency of dispersion. Zone 2 must be kept at the temperature of the melt in this region to prevent adverse heat transfer. For similar reasons and to prevent build-up due to freezing the discharge housing temperature must be kept close to that of the compound. The temperature of the screw must not be too high or it will cause premature melting in the blending compartment. On the other hand if the screw is too cold it can be shown that some materials will solidify on the screw and so reduce the working volume. Use of a die is optional with the MP mixer since the discharge zone serves the same function and applies back pressure to assist compounding. However, a die can be useful to direct the extrudate away from the discharge. Heating of the die is less cumbersome if a band heater is employed.

If the extruder has been left between batches with epoxy resin in the barrel then it is possible that overloading can occur when the screws are started at low speed. This can be due to unmelted resin and the starting operation should be repeated after a few minutes. Overloading due to cured material can be caused by inadequate cleaning between shifts. If the barrel contains cured material it will be necessary to open the barrel and clean thoroughly.

A slow screw speed of between 50 and 100 r.p.m. and a torque reading below 25% should be obtained before starting feed addition. The variable orifice should be partially closed to maintain a back pressure until stable compounding conditions are achieved, when the orifice may be opened more fully. This adjustment may be made without interrupting the compounding process.

When the feed rate and the screw speed are stabilized at around 90% torque the compounding process will require little attention from the operator.

To close down the machine, the feed must first be discontinued and the extruder allowed to run until discharge has almost ceased. It is recommended that small amounts of granulated resin are carefully added to purge the barrel and screws.

When the colour has to be changed, the cleaning procedure, mentioned later in this chapter should be carried out.

The torque reading should drop below 15% before reducing speed to the minimum and closing down.

The die head and discharge housing should be disconnected for cleaning well before start up on the next shift so that any build-up of partially cured product may be removed. A section of the screw operating within the discharge housing may be removed for overnight immersion in strong solvent or immediately replaced with a clean set if a colour change occurs during a working shift.

Advantages of the MP mixer

1. It is not excessively priced.
2. Relatively low power consumption.
3. Twin screw dispersion is more efficient than single screw dispersion and smaller amounts of strong colour are required due to the efficient colour development.
4. Maximum work load can be obtained.
5. It is easy to clean.
6. The dwell time of the melt-mixed material is short so that systems of high reactivity can be compounded.
7. The overload cut out protects the machine against accidental damage.
8. The screws may be removed by disconnecting a rigid coupling so that it is unnecessary to dismantle the gear box to remove and replace the screws.
9. It occupies a relatively small floor area.
10. Output rates are high. A two-inch machine produces more than 100 kg. per hour while a four-inch machine can approach an output of 1000 kg. per hour.
11. Finely powdered materials can be fed through the twin screw machine.

Disadvantages

1. If the feed hopper is not maintained at a fairly constant level there is a tendency to surge due to demixing of the feed stock which can cause overloading.
2. When the machine cuts out during the running period it may be impossible to restart without opening the barrel to remove accumulated feed. This fault could be avoided if the overload control would also cut out the vibratory feeder.

A refinement of the MP mixer is the twin screw extruder supplied by Werner and Pfleiderer under the product reference ZDS-K.

The processing section consists of two intermeshing screws, rotating in the same direction and at the same speed. The screws are self wiping and form a "sealing profile" through which dead spaces are minimized and uniform stress is ensured.

Fig. 33. Sealing profile of screw elements and kneading blocks (Courtesy Werner & Pfleiderer A.G.)

Fig. 34. Screw elements, kneading blocks and completed screws (Courtesy Werner & Pfleiderer A.G.)

The advantage over the Baker Perkins machine is the modular or building block principle for the variable design of the screws and barrel. Screw elements consist of different lengths and pitches with special kneading elements of various widths which are interchangeable. These elements and blocks are prevented from rotating on the shaft by a key running its full length. In the axial direction the configuration relates to the screw tip at the end of the shaft. The screw configuration can be changed easily to impart different shear conditions to the mix. Back pressure may therefore be increased by screw geometry rather than movement of screw relative to the discharge section.

The barrel also consists of individual sections.

ANCILLARY EQUIPMENT

Feed Hoppers

The single screw extruders are best fed from a direct feed hopper tightly seated in the feed inlet. Size segregation will occur during the filling of the hopper and this will cause the fine particles to accumulate at the centre. When the hopper is agitated or vibrated, large granules will rise to the surface. The rate of discharge of powdered or granulated materials is independent of the head of material in the hopper unless this head is close to the point of discharge. Most of the powder weight is carried by the walls of the vessel and unlike fluids the pressure caused by a powder does not increase linearly with depth but tends exponentially to a limiting value. It is therefore important to ensure that the rate of discharge from a hopper is maintained at a fairly constant value and this rate will only alter appreciably when the hopper is almost empty.

For continuous extruder production it is important to have level limit controls within the feed hopper.

The factor which will most affect the feed rate of premix is the manner in which the powder packs within the hopper. This will be dictated to a large extent by the method of pouring the premix into the hopper. To ensure that a tightly packed premix will flow, it is necessary to introduce slow moving blades within the hopper. This action will promote the dilatancy which is a necessary first step for the agglomerates to change shape and begin flowing.

While the action of the blades will assist segregation of particles it will prevent tight packing occurring.

Twin screw extruders require a controlled feed which is best achieved by a vibratory trough which is generally fed by a hopper without blades.

A premix with poor flow properties within the hopper will exhibit the effect shown in fig. 36 which is commonly referred to as "rat-holing".

A vibrating device fitted to the wall of the hopper will reduce this effect but a mechanically agitated hopper is preferable for use with a vibrating trough feed.

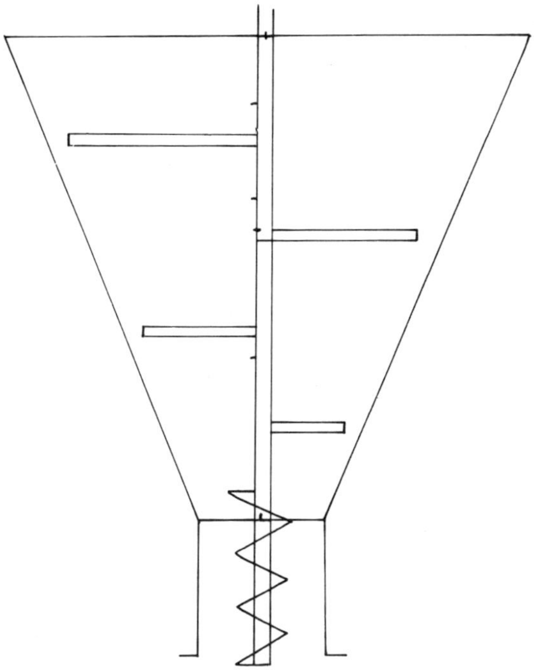

Fig. 35. Feed hopper with rotating blades

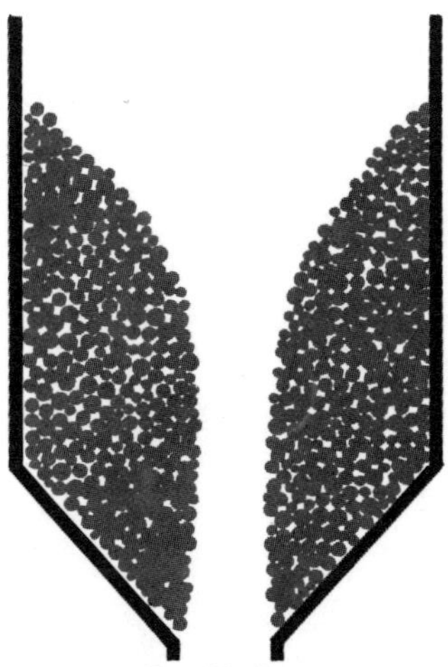

Fig. 36. Rat-holing

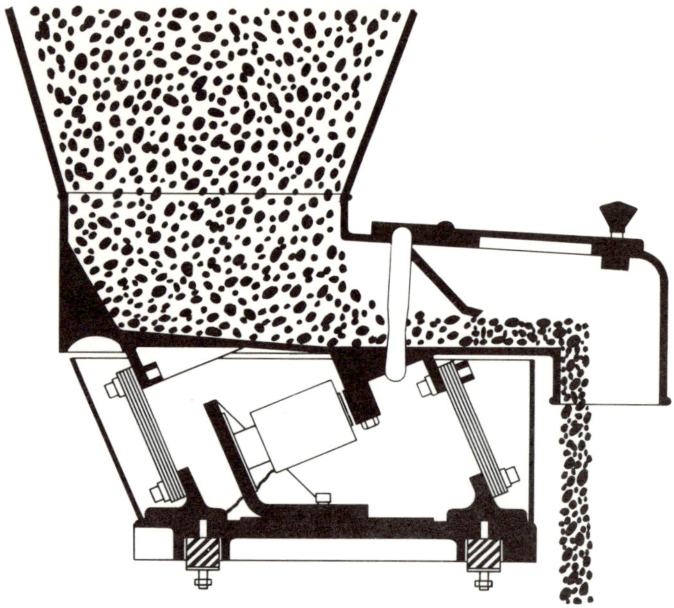

Fig. 37. Vibratory feed principle (Courtesy K.E.K. Ltd.)

Die-heads

A die is necessary with the Eickhoff extruder but it may be omitted from other extruders.

The die serves a twofold purpose: it produces back pressure in the extruder to assist compounding, and it directs the extrudate as a continuous ribbon to the cooling band.

Various different shapes of orifice have been assessed for suitability with thermoset powder coatings and it is generally agreed that a circular central orifice is preferable. There must be little dead-space between the orifice of the die and the end of the screw to prevent build up of cured extrudate. Inside the die the whole surface should be smooth and polished to facilitate cleaning. Mechanical chipping of cured extrudate must be avoided for this can easily damage the metal surface.

To assist the quick change of the extruder to a new colour it is recommended that two interchangeable dies are available so that the used die can be left to soak in strong solvent overnight, which will completely remove from it all cured extrudate.

If only one die is available and mechanical chipping is necessary to facilitate a quick turn round, then the operator is advised to use only copper or brass metal strips to avoid damaging the steel surface and to cool the die thoroughly by shock cooling in water to embrittle the built-up extrudate.

The die is best directly heated and cooled by water. It should be no more than a flat disc so that one man can handle it without discomfort when it is

Fig. 38. View of feeding device (Courtesy Sandvik Conveyors)

Fig. 39. Extrudate entering cooling roll (Courtesy Sandvik Conveyors)

hot. Water connections should be easily detachable. Electric band heaters do not provide very efficient temperature control but they have the advantage that they are easily removed for die-cleaning.

Cooling Rollers and Band

Most of the extruder manufacturers offer a cooling band for use with their own extruder and it is strongly recommended that this item should be included in the extruder order. The operator can then rest assured that the cooling band/rollers will be capable of handling the maximum output of the extruder.

Although most manufacturers of extruders will offer a cooling band to suit the output of their extruder, it may be more economical to purchase directly from a company specializing in this type of equipment.

Sandvik have world renowned experience in the manufacture of cooling conveying bands and they have designed suitable cooling bands for handling powder coating extrusions.

It has been found in practice that excess capacity for handling extrudate to the stage where it is a brittle, friable material suitable for immediate fine

Standard Cooler

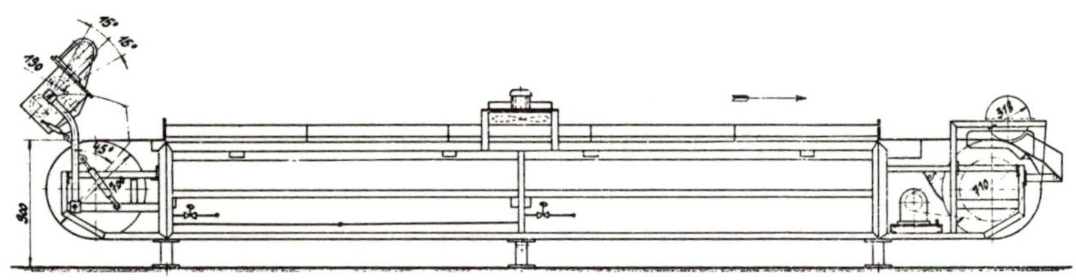

Table 14. Drawing of Standard Cooler with Technical Data and Dimensions
(Courtesy Sandvik Conveyors)

Plant Size / Title	Nominal capacity: Water cooling only	Nominal capacity: Water and air cooling	Belt-width and thickness	Max. product feeding width	Cooling zone: Length	Cooling zone: Cooling area	Cooling water consumption: Cooling tanks	Cooling water consumption: Feed rollers
Unit	kg/h	kg/h	mm	mm	m	m^2	m^3/h	m^3/h
1	100	125	400/0·8	280	2·5	0·7	0·65	0·75
2	200	250	400/0·8	280	5·0	1·4	1·3	0·75
3	350	425	600/0·8	480	5·0	2·4	1·3	1·2
4	525	650	600/0·8	480	7·5	3·6	2	1·2
5	750	900	800/0·8	680	7·5	5·1	4	1·5

grinding, is one of the most important rate determining stages in the continuous production of powder coatings.

Although the dwell time in the extruder is relatively small, there will be some partial reaction of the hardener especially with accelerated systems, and it is vitally important to check this reaction by rapid cooling of the extrudate. To maximise the cooling rate of the extrudate, the ribbon discharge from the die is fed between cooled steel rollers and squeezed into a wide continuous band of semi plastic material with a thickness of only two to three millimetres.

At high production rates the action of the cooling rolls is inadequate for complete cooling to friable material, and it is therefore necessary to continue the cooling process as the material is conveyed away from the cooling rolls, so that the material is completely cooled by the end of the cooled conveyor and can be easily broken up into flakes.

If the flaked extrudate is to be stored before fine grinding then it will not be necessary to cool immediately to ambient temperature and a less efficient cooling band may be employed. However, a band designed to be part of a continuous powder production line must be highly efficient, for the feed to the fine grinding machine must not be so warm as to cause melting in the grinding chamber of the mill.

Cleaning Procedures for the Extruder

When the production run is concluded the efficiency of cleaning necessary will be decided by the type and colour of the next batch to be manufactured.

In every case, however, a minimum level of cleanliness must be achieved, to ensure that trouble-free operation can be recommenced with the next batch. For closing down in the middle of a long run it will often be enough to purge the barrel well with epoxy resin. This material need not be wasted for in many cases it can be granulated and used in a further premix. The die will normally be removed for cleaning after the purging operation.

Cooling	Current supply			Dimensions			Total weight approx.
Air consumption	Belt drive	Roller drive	Air fan	Total length	Total width	Centre distance	
m^3/h	kW	kW	kW	mm	mm	mm	kg
1,200	0·75	0·75	0·1	4,850	920	3,800	820
1,200	0·75	0·75	0·1	7,350	920	6,300	1,050
2,000	1·1	1·1	0·25	7,350	1,120	6,300	1,650
2,000	1·1	1·1	0·25	9,850	1,120	8,800	1,880
2,000	1·1	1·1	0·25	9,850	1,320	8,880	2,400

To clean the barrel and screws more thoroughly the machine may first be purged with soft vinyl to scour the working surfaces, followed by a choice of absorbent soft materials which are first soaked in a high boiling point ketone or glycol ether solvent. Suitable materials are crepe paper, soft cardboard, soft polyurethane foam. After feeding these materials for some twenty or thirty minutes, the extruder may be shut down with solvent containing material left inside the barrel, or the barrel opened and the last traces wiped away with a solvent wetted rag. It is possible to recommence the run after the porous solvent containing material has been left in the barrel overnight, by simply purging with epoxy resin followed by premix.

References to the equipment described in Chapter 6 will be found in the section relating to suppliers.

7

Granulation and Fine Grinding Processes

SUMMARY

The fine grinding process describes the reduction of the particle size of the compounded powder to that required for application. Ease of particle size reduction will depend largely upon the molecular weight of the base resin. While thermosetting resins have relatively low molecular weights and are generally quite brittle at ambient temperatures, thermoplastic resins, typified by nylons, PVC and polyethylenes, are extremely tough and resilient.

High molecular weight materials are therefore reduced in size at low temperatures when they are more brittle and friable that is by freeze grinding or cryogenic methods.

The term "grinding" as used in the paint industry is really a misnomer since the processes used are merely for dispersing pigments that have already been ground and are supplied in fine particulate form. Paint "grinding" processes apply shear forces to break down agglomerates and cause the individual particles to be "wetted" by the dispersion media.

Fine grinding of solids is a process of attrition in which size reduction is effected by impacting the solid lumps against a grinding surface or by causing particles to be reduced in size by impacting with each other.

The utilization of energy within a high-speed grinding chamber is poor and the excess mechanical energy is largely converted to heat. The relatively low softening temperatures of powder coating resins make it important to keep mill temperatures close to ambient for maximum output. Close temperature control has a marked effect upon the throughput, the range of particle size, the particle shape and the ease of cleaning.

Several methods are used to control the temperature build up within the mill:

1. Air cooling is effective when a large air throughput is used to transport powder and to classify the product. This method is more effective with refrigerated air which may be essential during hot weather.
2. Water jackets may be employed to cool the static working faces of certain mills and this system is also improved by using conventional methods to cool the water.
3. The feed into the grinding chamber may be mixed with dry ice (solid carbon dioxide) produced by a snow machine, or liquid carbon dioxide may be injected into the feed hopper or grinding chamber.
4. Liquid nitrogen may also be injected into the feed hopper or grinding chamber.

High air flow is difficult to control because it is often related to an in-line classification system.

Production rates on air cooled grinding machines do suffer in hot weather and it is difficult to grind certain polyesters and acrylic powders and to refrigerate sufficiently well to show a marked improvement in the process of air cooling.

Water is an efficient cooling medium and it is not difficult to cool. However, the cooling area of the jacket is often quite inadequate to cope with the desired flow through the grinding chamber.

Water cooled mills generally have lower throughput rates than air cooled mills.

Dry ice is more effective than the first two methods but requires relatively large amounts of solid CO_2 to give effective cooling. It is difficult to maintain thermostatic control, and the large amounts of carbon dioxide involved can affect the eyes of the operators and their breathing.

Liquid nitrogen is the most favoured method for like CO_2 it is essentially non-toxic, it may be added in relatively small quantities which respond readily to thermostatic control, it also provides an inert gas atmosphere within the grinding chamber to reduce the explosion risk, it avoids the necessity for separate refrigeration devices and it allows both thermoplastic and thermoset polymers to be ground within the same mill. Liquid nitrogen is, however, relatively expensive and the cost of the liquid can add between 2 and 10% to the *total manufacturing costs.*

GRANULATION

Granulation of the compounded material is only necessary to ensure an acceptable feed size to the fine grinding mill. Extrusion compounded materials are invariably granulated by a flaking roller at the end of the cooling band. Material which has been allowed to cool in trays after Z-blade compounding is usually granulated with the aid of a machine operating the Masson cutter principle. These machines have a deep inlet throat leading to a cylindrical chamber where robust rotating knives quickly reduce the lumps to granules which fall through a large aperture screen between the rotating knives.

These devices are essential for handling tough and resilient thermoplastic resins; brittle thermosetting products present no problems of granulation. In fact a relatively small machine can handle several hundred kilos of compounded thermoset material per hour.

Another machine for granulating thermoset powders is the kibbler.

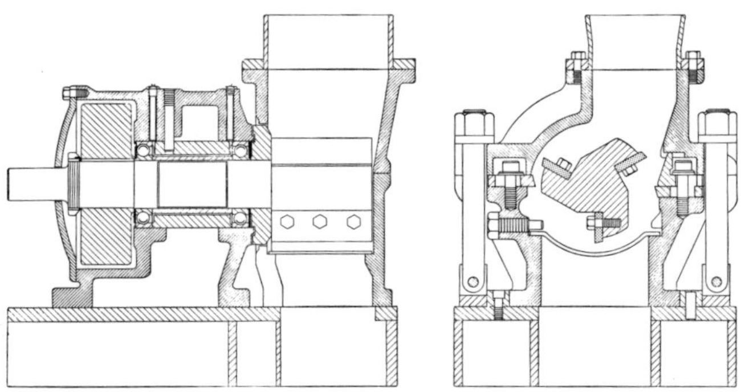

Fig. 40. Sectional drawing of U.S.I. Cumberland granulator (Courtesy U.S.I. Ltd.)

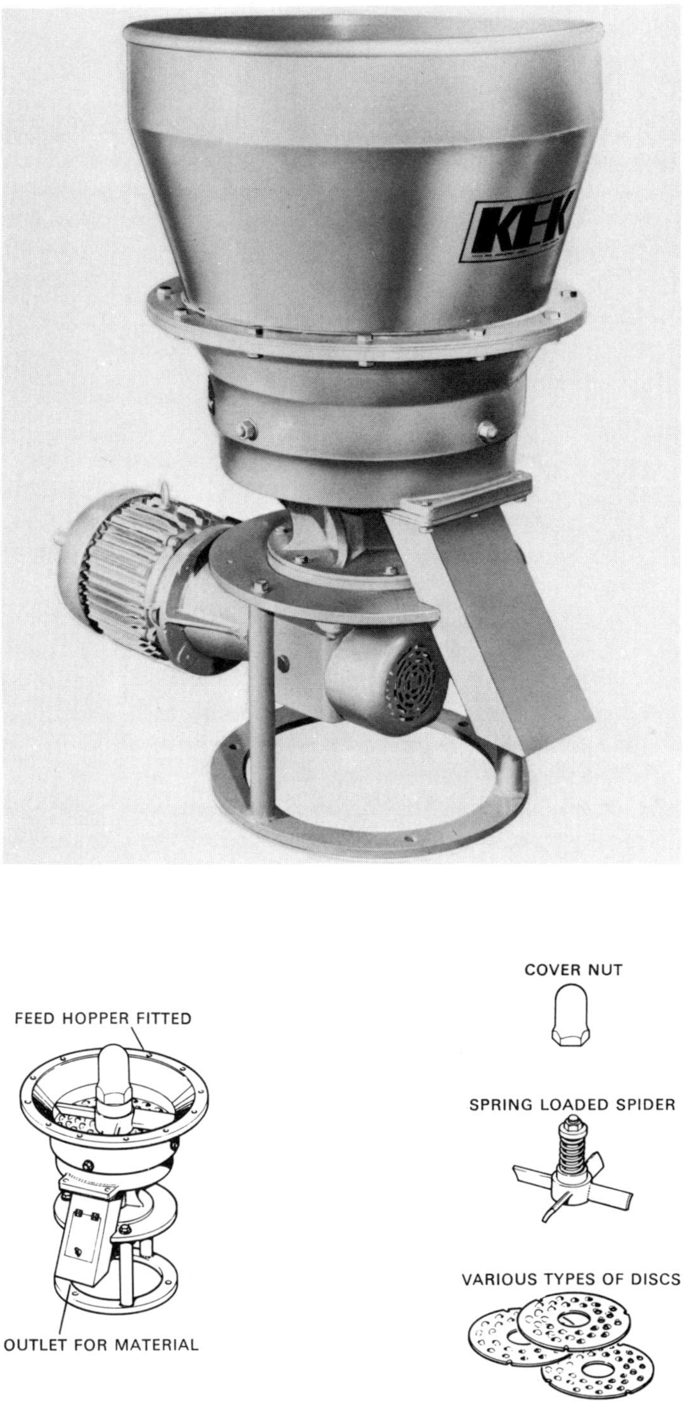

Fig. 41. Kibbler (Courtesy K.E.K. Ltd.)

FINE GRINDING

THE HAMMER MILL

The grinding action takes place by impact between the rapidly moving hammers and the granulated feed. In the grinding chamber there is a variable speed rotor fitted with hammers, a cover with a multiple deflector lining, a retaining screen at the mill exit and a feed screw to ensure uniform delivery of the granulated powder.

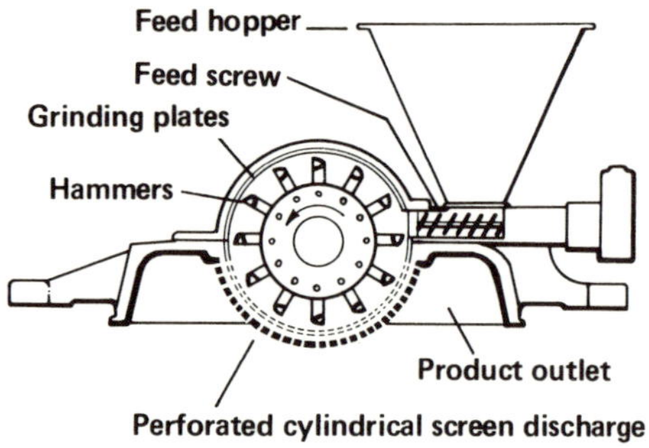

Fig. 42. Hammer mill (Courtesy Mikropul-Ducon Ltd.)

The three main factors which affect the particle size are the type of hammers used, the speed of the rotor and the size of the screen.

There are two types of hammer; the stirrup type and the bar type. For fine particle size distributions the stirrup is recommended while the bar type produces a minimum of fines. The working edges of the hammers are tipped with an abrasion resistant alloy.

A coarse grind will result from a low rotor speed, higher speeds produce finer particles.

The deflector liner is replaceable and the serations provide ledges where the moving particles are further reduced in size due to impact with the liner.

Retaining screens may be round perforated, herringbone slotted or cross slotted. The round perforated screen is structurally stronger but the other two varieties are preferred for grinding powder coatings since they both give a finer particle size. If the powder coating is required to be sufficiently coarse for fluidized bed application then a jump gap screen is fitted. This comprises a series of bars which are so sloped to throw the particle back into the mill until it is sufficiently reduced in size to pass through the gap.

Suitable hammer mills for grinding powder coatings are supplied by Mikropul Ltd., Alpine (Machinery) Ltd. and Hosakawa.

Advantages of Hammer Mills

1. Relatively inexpensive.
2. Do not occupy large floor areas.
3. Extremely robust for continuous operation.

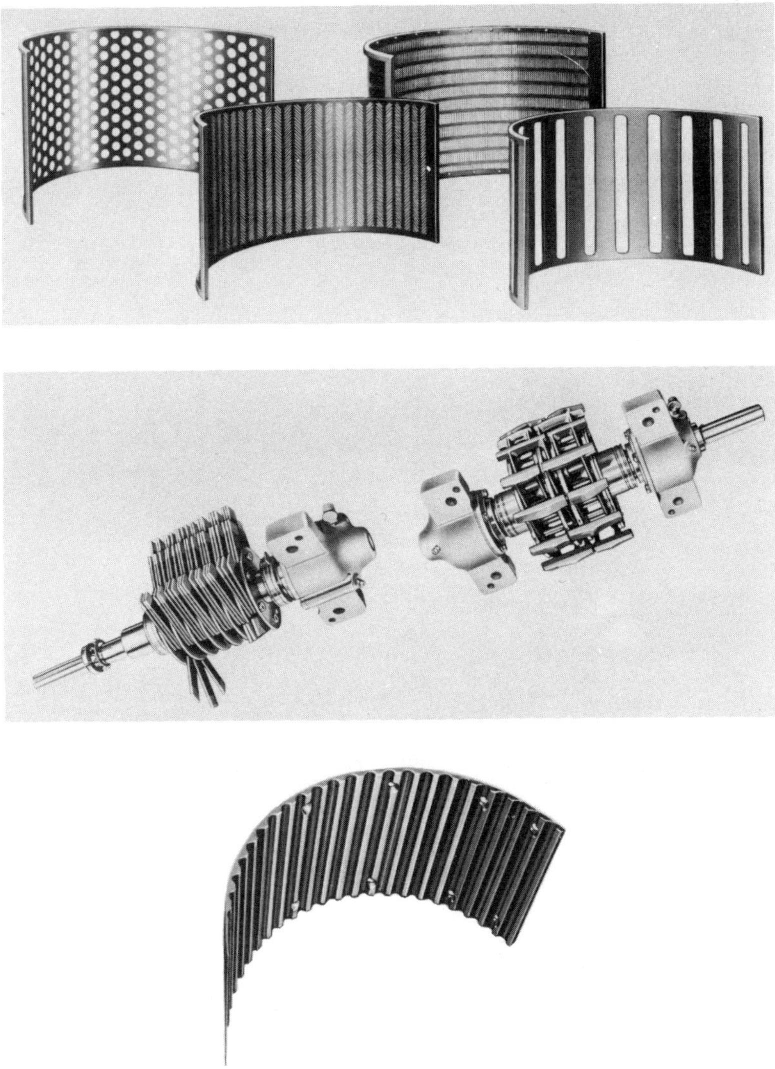

Fig. 43. Screens, hammers and liners (Courtesy Mikropul-Ducon Ltd.)

4. Useful for grinding powders to the fluidized-bed range of particle sizes.
5. Easy to clean.
6. Laboratory models exist for small quantities.

Disadvantages

1. Noisy in operation.
2. Do not air classify.
3. It is not easy to cool mills and melts occur frequently during warm weather.

THE CROSS FLOW MILL

The cross-flow principle is illustrated by the Alpine Ultraplex mill. The term "cross flow" derives from the directional movement of the material within the grinding chamber between the beater mechanism and the baffle-ribbed grinding track. Retention time in the mill is determined by the slanted, baffled ribs so that particles are prevented from escaping crosswise to the

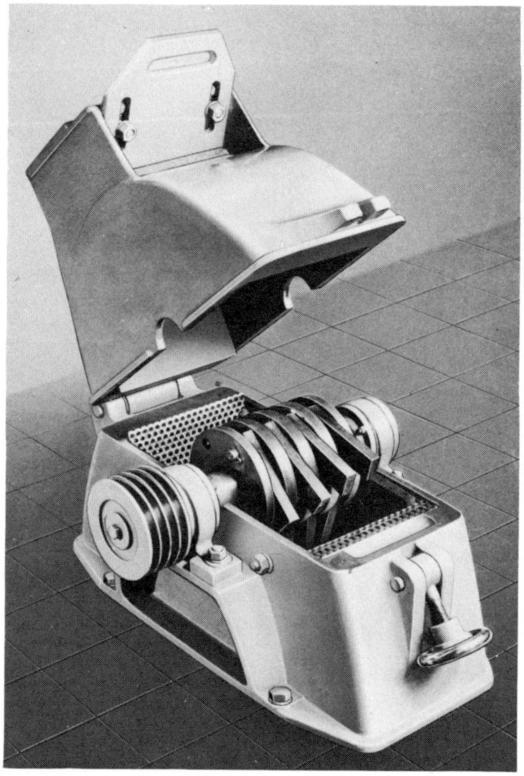

Fig. 44. Omniplex hammer mill (Courtesy Alpine (Machinery) Ltd.)

direction of impact or by following the air currents, until the friction drag of the air overcomes the kinetic energy of impacting.

Feeding is mainly carried out within a vibratory channel or from a vibratory feed hopper. The granulates enter the central section of the mill and the particles are flung against the baffles by the beater motion.

An annular gap near the rear wall of the housing provides the fine material exit.

Fig. 45. Ultraplex cross flow mill (Courtesy Alpine (Machinery) Ltd.)

Advantages of the Cross Flow Mill

1. It will classify the powder to ensure that no coarse particles escape from the mill.
2. The collection hopper is relatively small.
3. The grinding chamber is easily accessible and therefore it is easy to clean,
4. Relatively inexpensive.

Disadvantages

1. It is noisy in operation.
2. The mill is difficult to cool and trouble can occur with melting in warm weather.

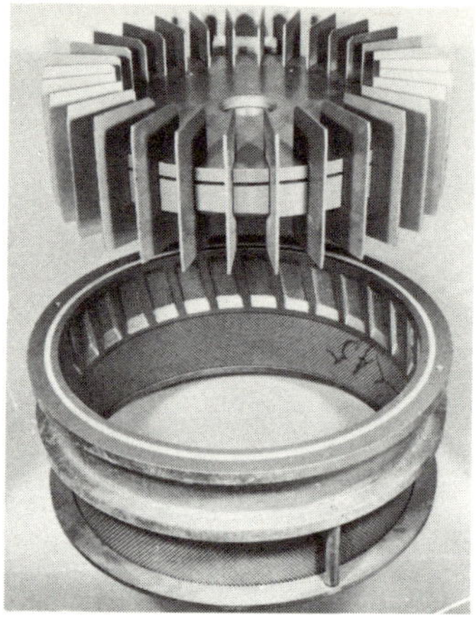

Fig. 46. Beater mechanism and screens for ultraplex mill (Courtesy Alpine (Machinery) Ltd.)

THE PIN DISC MILL

The principle is used in the design of the range of K.E.K. fine grinders and in the Alpine Kolloplex and Contraplex Mills.

These are sieveless stud mills with one or two rotating stud or pin discs. Control of the particle size is effected by adjusting the speed of the rotor and by increasing or decreasing the number of studs.

Fig. 47. Contraplex wide chamber mill 400 (Courtesy Alpine (Machinery) Ltd.)

Size reduction occurs by the impact of the particle at the pins of the intermeshing rows of pin discs. Feed is introduced centrally drawn in by the air intake flow produced by the rotating pins. As each coarse particle is flung out centrifugally between the pin disc, it is reduced to a number of fine particles by impact. Powder is prevented from reaching the bearings of the mill by the action of ventilator fans behind the pin discs which causes air to be blown into the mill via a ring gap. This gap can be adjusted to improve air cooling.

The K.E.K. mill allows the pins to be replaced by a turbine grinding wheel and screens for handling less friable compounds.

Advantages of the Pin Disc Mill

1. Sieving is generally not necessary and screens are not employed within the mill.

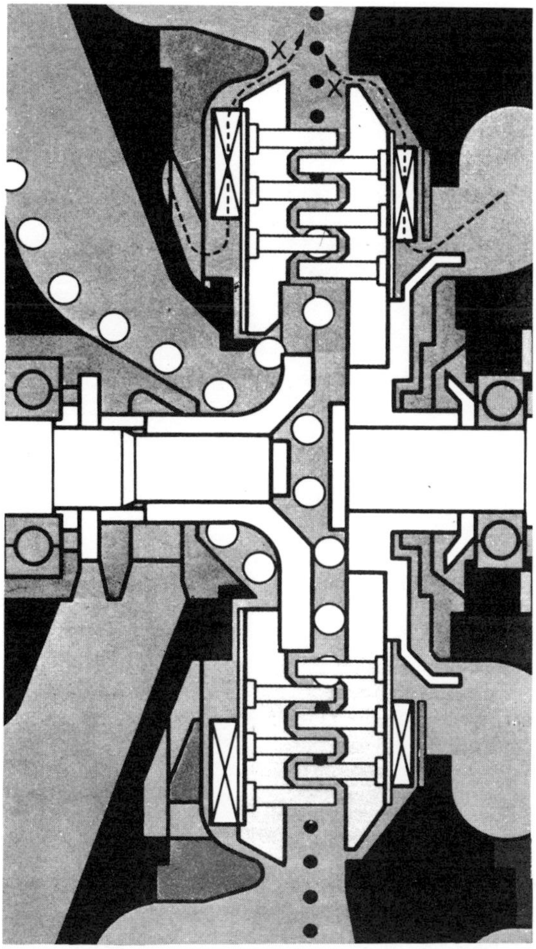

Fig. 48. Cross section of Contraplex mill (Courtesy Alpine (Machinery) Ltd.)

2. Relatively inexpensive.
3. Occupies little floor area.
4. It is easy to clean and maintain.

Disadvantages

1. Good temperature control is difficult in warm weather and pins are difficult to clean when a melt occurs.
2. It is difficult to avoid excessive amounts of fine particles.
3. It is a noisy machine.

Fig. 49. 4H mill and turbine wheel (Courtesy K.E.K. Ltd.)

THE MIKRO ACM PULVERIZER

This series of fine grinding machines is supplied by Mikropul Ltd. and marketed by a worldwide network of associated companies. There are three sizes of grinding mill, the 10, 30 and 60 ACM, with the numbers relating to the horse power rating of the pin rotor drive.

Fig. 50. The Mikro ACM mill (Courtesy Mikropul-Ducon Ltd.)

Granulated extrudates are fed into the grinding chamber by a variable speed screw where the granules encounter a set of pins which can be rotated at speeds ranging from 4,500 to 8,500 r.p.m. Fine grinding results from the attrition of the granules on the hardened serated lining of the mill. Particles are conveyed upwards by the air stream which enters below the pin table, and they are deflected downwards by baffles to continue the grinding process until they are fine enough to be entrained by the air stream. Air leaves the mill through a rotary vane, the rotational speed of which may be adjusted to between 600 and 2,000 r.p.m. to control the size of the particle leaving the mill. Each aperture of the vane serves as an exit door for the particle to leave the mill. The faster the vane rotates the smaller is the exit aperture and the particle must therefore become smaller before it can escape. This selective choice of rotary vane speed and the control it exerts upon the fineness of grind is termed classification. While it is possible to ensure that the bulk of the powder particles will lie below an upper limit, in the region of 80 microns, it will not prevent the formation of excessively fine particles. The fines may be removed by conveying the powder/air stream through a second, in-line classifier in series which permits only the fines smaller than 10–15 microns to pass through for collection. Suitable adjustment of the two in-series classifiers can result in a close control of particle size.

The second in-line classifier is not a standard feature of the ACM mill and it can only be installed in a standard production model if the air stream can be considerably increased with corresponding increase in the filter area of the bag collectors.

In the majority of electrostatic spray applications of powder, it will not be necessary to remove the fines. In large plants where oversprayed powder is collected and re-circulated, the removal of fines may be important to prevent blockage of injectors and build-up of powder along narrow pipe lines.

Collection and Filtration of Dust

Collection of powder is a most important feature of these mills for the high output requires large scale, efficient, dust control at the collection point.

The principle requirement of filter bag collection is a large enough collection area to permit the air in the powder/air stream to be discharged through the filter bag to atmosphere without producing a back pressure which would affect the efficiency of the milling operation. Fine powder particles are retained on the filter bed and this accumulation must be dislodged to prevent the powder blocking the pores of the filter medium.

Woven cloth has been found to be unsuitable as a filter for it is possible that fine powder can seep through open pores. There is also a pronounced tendency for dust particles to bridge across the pores and reduce the efficiency of filtration (see fig. 51).

The heavily matted texture of felt filters is shown in fig. 51 and even submicron particles are unable to penetrate the dense body of the cloth. When this material is used with reverse air jet cleaning a high efficiency of air filtration results.

Efficiency of filtration through the filter cloth is improved by chemical pretreatment of the material to reduce the possibility of seepage of fines.

The area of filter surface will depend entirely upon the volume of air being expelled by the mill and as a general principle the ratio of the filter area in square feet to the volume of air in cubic feet per minute is approximately 1:15.

Collection units are, therefore, quite large and cumbersome compared to the size of the grinding mill.

The most common filter is a cylindrical felt bag which may be attached to collars in the steel housing of the collector unit. Fine powders may be dislodged from the filter surface by

(a) Simple mechanical shaking of the bags.
(b) Intermittent reverse air jets of high pressure.
(c) Continuous traversing of the cylinder wall with a blow ring.

While system (a) is suitable for the filter beds of grinding machines with low throughput rates it cannot be considered for use with a high-throughput continuous grinding process.

The number of filter bags necessary to give adequate filtration area, and the ease of accessibility, will be important factors to bear in mind if the grinding machine and collection unit are to be used with a wide range of different colours and types of powder.

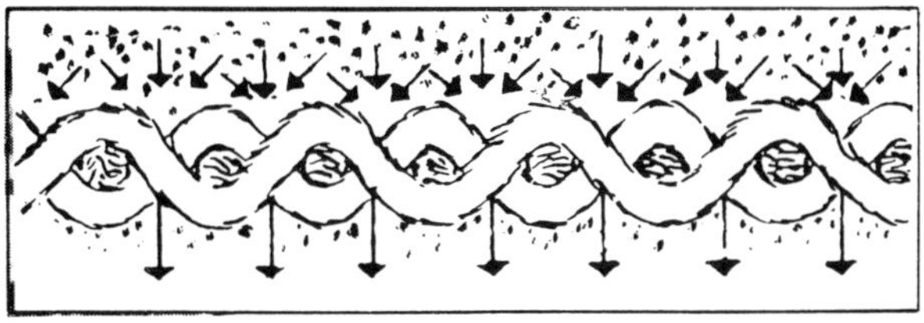

WOVEN CLOTH

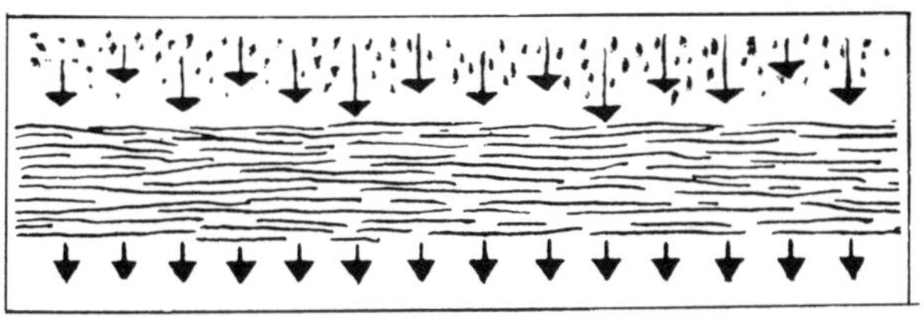

FELT

Fig. 51. Filtration principle (Courtesy Mikropul-Ducon Ltd.)

Cleaning the filter bags is a time consuming operation, and a sixteen bag filter unit can only be used, when continual colour changing is required, if the whole unit can be removed and simply replaced by a separate unit.

The blow-ring operated filter bags are easier to clean and replace than the filter elements used in the reverse high pressure, air jet process described in the Mikro-Pulsaire operation.

The Mikro-Pulsaire consists, primarily, of a series of cylindrical filter elements enclosed in a dust-proof housing. Dust laden air is admitted to the housing, and clean air is withdrawn through the filter cylinders. As dust laden particles accumulate on the filter elements, periodic cleaning is achieved by the introduction of a momentary jet of compressed air through a specially designed venturi, mounted above each filter cylinder. The compressed air jet pumps secondary air, producing a reverse air flow sufficient to clean the filter cylinders. Since only a small proportion of the total filter area is cleaned at one time, continuous flow through the collector is maintained. The jets are controlled by diaphragm valves, activated by solenoid pilot valves and a timer.

While the dust is allowed to collect on the outside of the filter bags used with the Mikro-Pulsaire, the filter bags cleaned by blow-ring methods allow

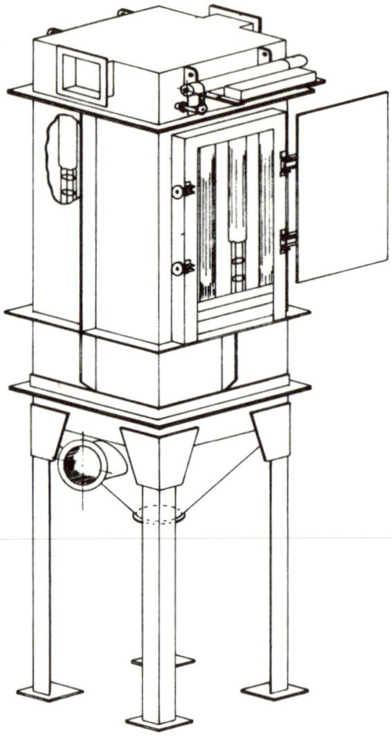

Fig. 52. Mikro Pulsaire collector Type 16S-6-30 (Courtesy Mikropul-Ducon Ltd.)

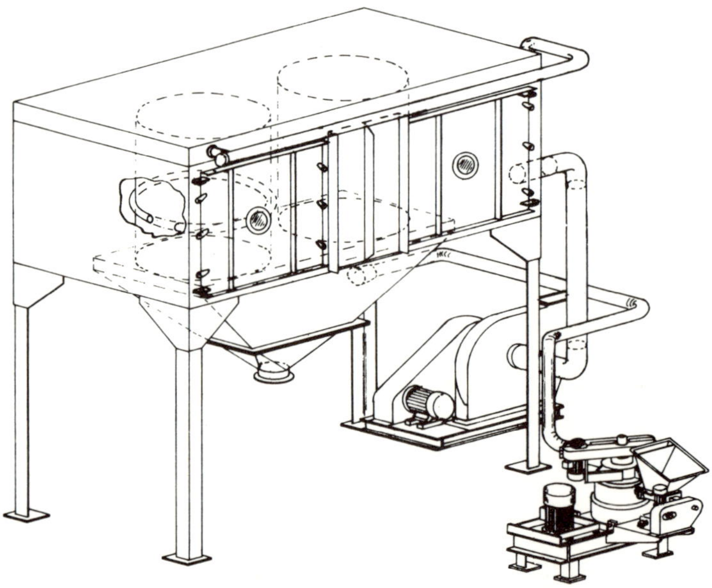

Fig. 53. Blow ring collector Type 36-2-42 with 10 ACM Mikro pulverizer (Courtesy Mikropul-Ducon Ltd.)

the powder to collect inside the filter bags. The latter method obviates the need to support the filter bags with a cylindrical wire frame.

The dust collected inside the blow-ring cleaned filter bags is dislodged by a jet of high velocity air from a slotted orifice located in the inner periphery of the blow-ring as it traverses the length of the filter bag. This air jet, combined with the flexing of the unsupported bag by the movement of the blow-ring, effectively dislodges the deposit of dust and allows it to fall into the collecting hopper where it is removed continuously by an air tight rotary valve.

THE FINE VICTORY MILL

An alternative to the Mikropul range of screenless fine grinding mills is offered by the well known Japanese firm, Hosokawa who have been manufacturers of powder handling machinery for more than fifty years.

The Fine Victory Mill also has a built-in classifier within the milling chamber. Granulated material is fed to the centre of the rotor where it is accelerated radially by the centrifugal force of the rotor blades. Particle size reduction occurs by attrition of the material between the rotor and the stationary liners sited on the walls of the milling chamber.

The finely ground material now enters the classifying chamber where the coarse particles are returned through an adjustable opening while the fines are swept away through the mill outlet.

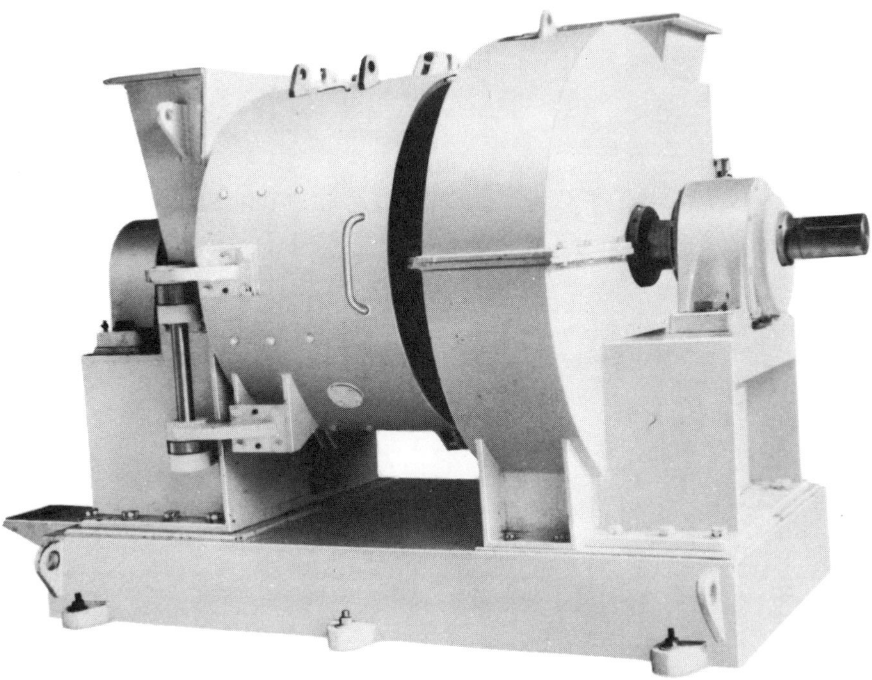

Fig. 54 Fine Victory mill (Courtesy Hosokawa Ltd.)

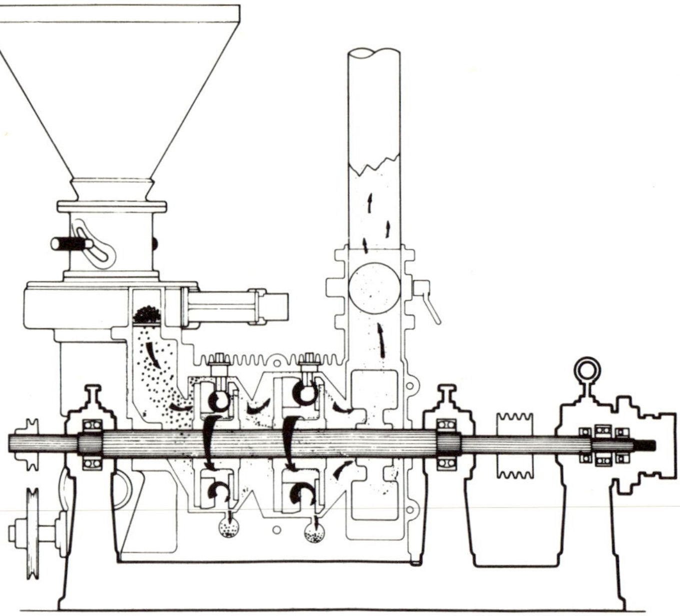

Fig. 55. Operating principle of the Fine Victory mill (Courtesy Hosokawa Ltd.)

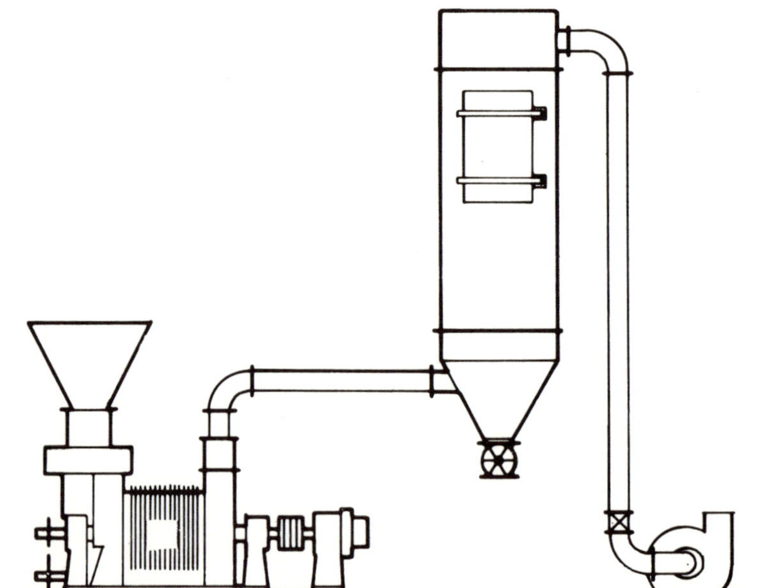

Fig. 56. Diagram of Mill feeding directly to the Pulsaire collector (Courtesy Hosokawa Ltd.)

An external operating control is provided to adjust the opening within the classifying chamber.

The mill can be constructed from mild steel or stainless steel and the impeller blades may be faced with tungsten carbide.

A mill of stainless steel construction would be suitable for cryogenic grinding and such a mill would have the capability of being used universally for both thermoplastic, high molecular weight products required for fluidized bed or electrostatic spray application, and the lower molecular weight thermoset powders mainly required for electrostatic spray.

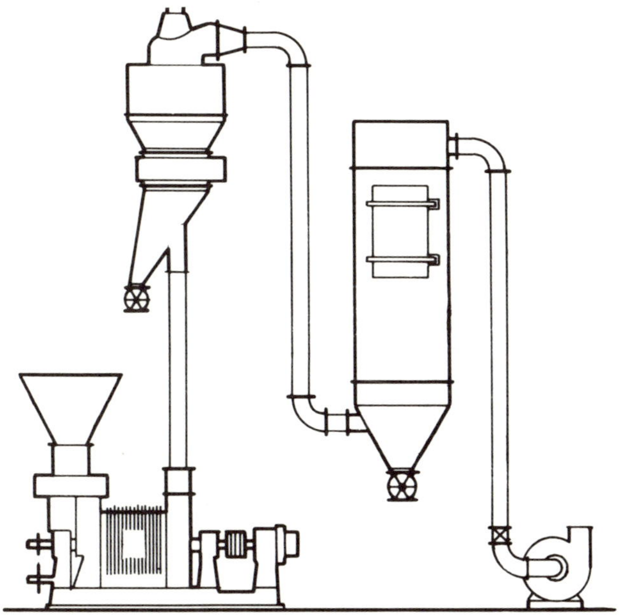

Fig. 57. Diagram of mill feeding to Micron separator thence to Pulsaire collector (Courtesy Hosokawa Ltd.)

There are five models currently available with outputs directly related to the h.p. rating of the main drive.

FVP 1	7·5 h.p.
FVP 2	15 h.p.
FVP 3	30 h.p.
FVP 4	50 h.p.
FVP 5	100 h.p.

The Fine Victory Mill can be used in conjunction with the Mikropulsaire collection units or the Micron Separator and Pulsaire collector.

THE FLUID ENERGY MILL

This method is included for the sake of completeness but it is seldom used for the grinding of thermoset powder coatings. The efficiency of milling is excellent and the majority of particles produced are of submicron size. There are no commercial methods of applying submicron powders at the present time, but there will be considerable investigations into fluid energy milling of powders due to the mounting interest in thin film powder coatings.

The term fluid energy relates to the use of fine high velocity gas streams which impinge at an angle upon the particles as they are entrained in the venturi shaped grinding and classifying chamber.

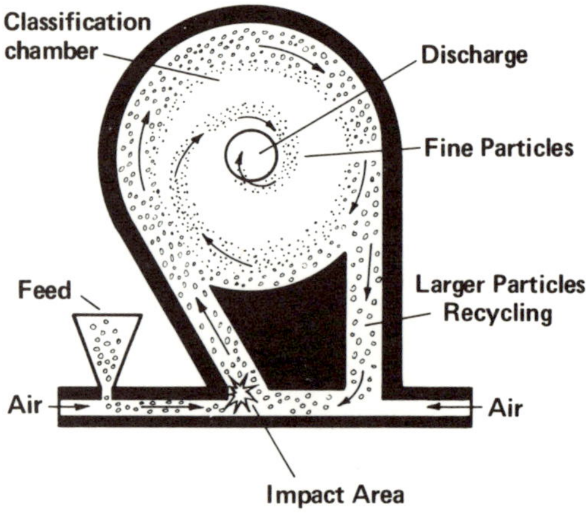

Fig. 58. Fluid energy mill

Design of the chamber is a most important feature for the particle must not be comminuted by attrition with the wall of the chamber, since the high velocities would cause considerable wear even with high strength steel. Size reduction is effected by direct attrition of the particles as they impact upon other particles. When the particle is sufficiently fine it passes out of the mill through a static classifier outlet. There are no moving parts in this mill which is relatively inexpensive. The major cost contribution arises from the need to provide high volumes of gas (usually air) at pressures in excess of 100 p.s.i.

THE CONTINUOUS BALL MILL

A relatively simple method for the fine grinding of powders is the continuous ball mill.

Advantages

The low energy used in the attrition process which reduces the percentage of fine particles; it requires little supervision; it does not generate excessive heat.

Disadvantages

The low throughput rate and the considerable wear on the glands of the axially sited inlet. Sieving is essential to remove the very high proportion of coarse particles which must then be recycled.

8

The Powder Particle

SUMMARY

The distribution of particle size in powder coatings is perhaps the most important characteristic discussed in this section, for this will considerably affect the application properties of both fluidized bed and electrostatically sprayed powders.

Handling properties and the ability of the powder to exhibit fluid flow are mainly dictated by size distribution, although particle shape is a contributory factor. This comment is also applicable to the electrostatic charging properties. The current investigations into thin film applications for powder coatings have shown that particle size distribution can affect deposition efficiency, film flow, pinholing, and there can be little doubt that excessive amounts of fines in powders, below 7 μ, do create a health hazard if they are inhaled.

Classification is a means of separating powders into two or more fractions, each fraction having a relatively small range of particle sizes; the commercial methods for classification are discussed in this chapter.

The measurement of particle size by methods which are not time consuming, is an important subject which is also reviewed.

Certain equations can be derived to predict the behaviour of particulate solids during separation techniques and, in a later chapter, in the acceptance of electrostatic charge. It is assumed for the purpose of these predictions that particles are essentially spherical; however photographs of magnified powder particles illustrate the irregular shapes obtained in practice, so that the equations can therefore only be an approximation to true behaviour.

The rheological properties of powders composed of irregular shaped particles, ranging from under one micron to several hundred microns in "diameter", have proved difficult to measure in many industries because there is no simple test equipment for determining the flow characteristics of powders.

It is difficult to apply mathematical principles to the flow properties of liquids, but the behaviour of the solid/gas interface is even more difficult to assess, since the gaseous continuous phase does not assist the transmission of forces.

Measurement of the flow characteristics of particles through channels or orifices can help to solve storage and recovery problems. However, the only worthwhile observation is whether or not the particles are free-flowing.

The factors which affect the flow properties of powders may be listed as: particle size; particle shape; cohesive forces; adhesive forces; the presence of moisture; size segregation; electrostatic charge; density; the presence of flow aids, and packing of the particles.

A powder is conventionally described as a mixture of particles whose sizes range from 0 to 100 microns in diameter; this text-book definition can be applied to the mixture of particles comprising the majority of electrostatic powder coatings, these coatings can truly be called powders.

When particles range in size from 100 to 3,000 microns, such particulate materials are referred to as granular solids. Some coating materials, suitable for fluidized bed and flame spraying applications, have particles with diameters in this size range although the diameters of the largest particles are unlikely to exceed 500 microns.

Powders are said to be ultrafine when particle diameters lie in the range 0·1–1·0 microns, superfine when diameters lie in the range 1–10 microns and granular between 10–100 microns.

Particle size distributions of powders may be roughly assessed by sieves which can retain particles as small as 50 microns although separations down to 20 microns may be achieved with an air-jet sieve.

The mean projected area diameter is the diameter of a circle which is considered to have the same area as the particle at rest in a stable position.

Projected area diameter is larger than the sieve size.

Terms used to describe the powder state are, fractional solids content and voidage.

The fractional solids content is the ratio of apparent powder density to the effective particle density (or bulk density to solids density).

Voidage fraction is unity minus the fractional solids content.

A loosely packed powder will have a fractional solids content of approximately 0·60 while a settled and partially packed powder will have a fractional solids content of 0·65 or greater.

Cohesion describes the sticking of powder particles to each other while adhesion refers to the sticking of particles to a vessel wall or substratum.

Angle of repose is the angle to the horizontal assumed by the free surface of a heap of powder at rest. This simple measurement is often used to describe the relative flow characteristics of different powders.

The inherently brittle nature of thermosetting powders produces irregular shaped particles which do not pack easily when deposited as a powder film.

Acceptance of electrostatic charge by these irregular shaped particles results in the concentration of charge at the peaks and corners hence the behaviour of such particles in an electrostatic field is far from uniform.

Size segregation of powders or granular materials during storage etc, presents problems in handling heterogeneous mixtures at the preblending and extruder feeding operations. This problem has been discussed in a previous section. Analysis of particle sizes emerging from a vibratory feed hopper shows an excess of fines at the beginning of the feed due to percolation of fines to the bottom of the container. Similarly, at the end of the feed, there is an excess of coarse particles.

The several comments relating to size distribution in this summary should serve to underline the basic objective, to obtain a uniform particle size.

PARTICLE SIZE CLASSIFICATION

The existence of a range of particle sizes (a size distribution) is illustrated by two photographs. Figs 60 and 61. In the first of these the pigment titanium dioxide (Tioxide R-CR2) is shown at a magnification of 20,000. While the particles are extremely small by comparison with powder coatings it will be seen that the pigment particles are essentially uniform in size and regular in shape.

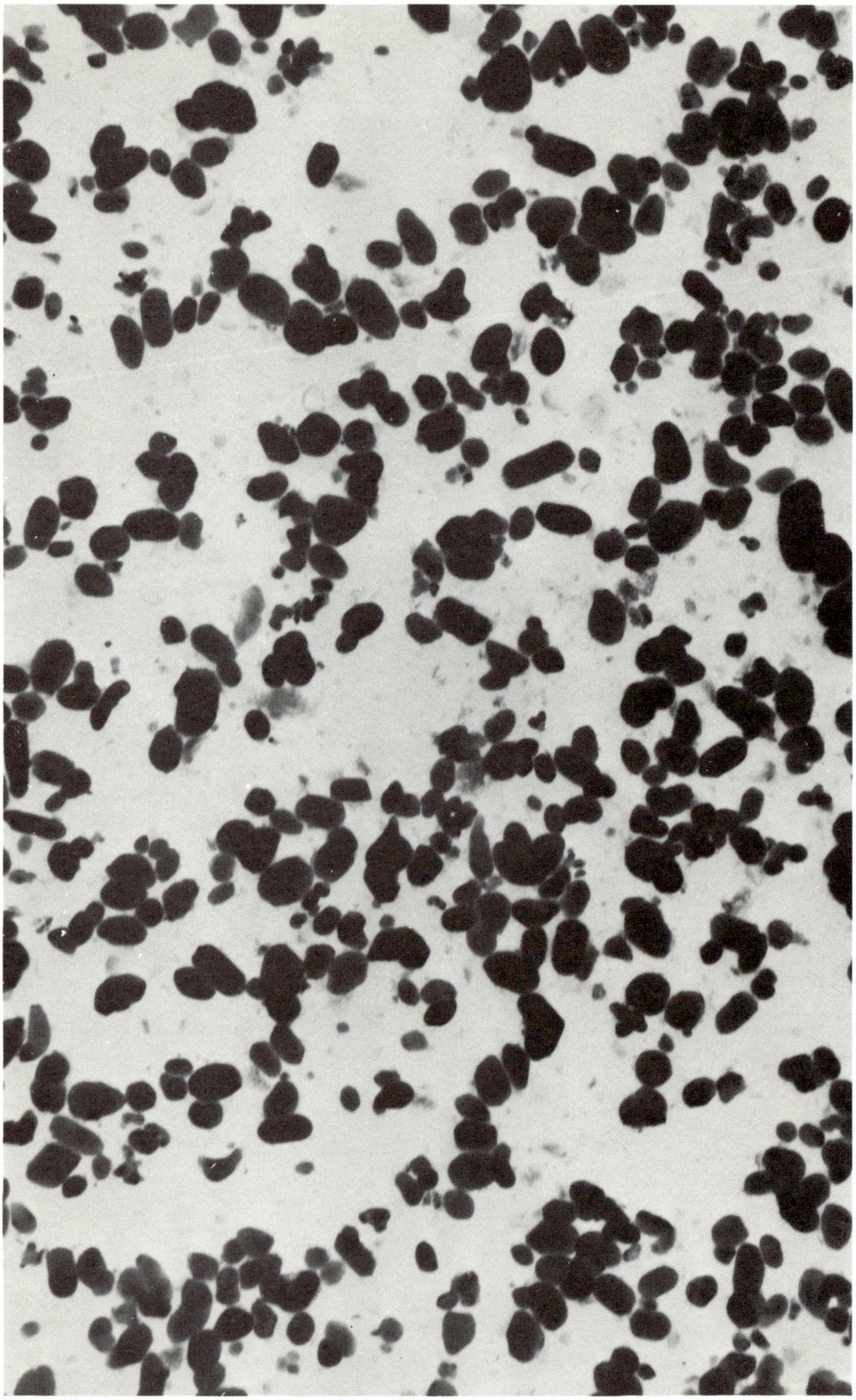

Fig. 59. Titanium dioxide (magnification 20,000) (Courtesy Tioxide International Ltd.)

The second photograph is a magnification by 100 of an epoxy powder coating, pigmented with Tioxide R-CR2, ground on a pin disc mill.

At higher magnification the irregular shape of the particles of powder coating is even more apparent.

The best known type of commercial classifier is the screen and there are several types which are currently in use.

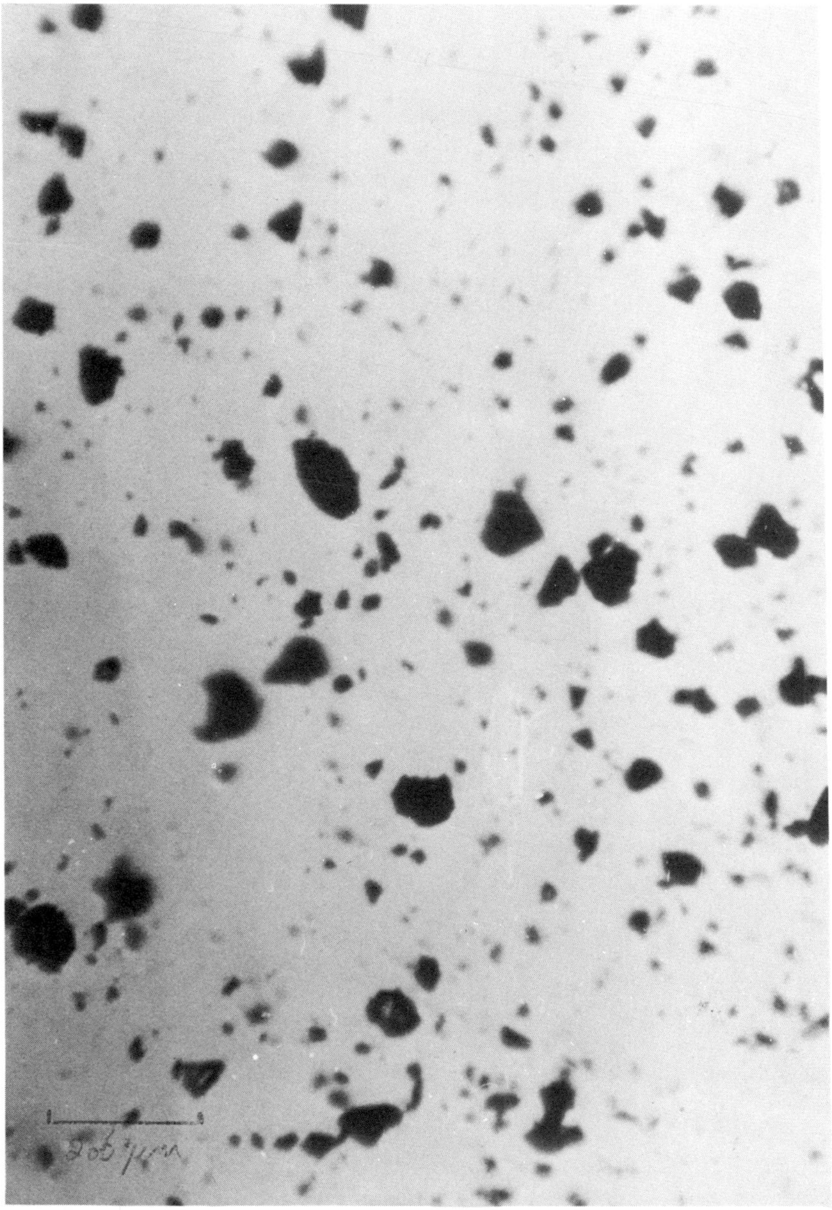

Fig. 60. Epoxy powder coating (magnification 100) (Courtesy Tioxide International Ltd.)

The Sifter

This is a simple machine, of robust construction, which classifies the powder by vibrating the fine portion through nylon or stainless steel meshes of chosen sizes.

Coarse material retained on the screen may be recycled through the fine grinding machine.

Fig. 61. The "Olympia" sifter (Courtesy K.E.K. Ltd.)

The vibratory action of the sifter is produced by a motor-driven cam and follower, which imparts a high-speed oscillatory action in the horizontal plane.

Rubber mountings support the sieving unit which is easily detached for cleaning. The material passing through the sieve collects in a pan beneath the sieve and is continually discharged via a chute.

Intensity of vibration may be adjusted by means of a rotating control.

While the sifter is relatively inexpensive, the powder readily blocks the screen, it is noisy and very dusty in operation. Certain cheap sifter models need to be anchored down to prevent them "walking away".

The Air Suspension Sieve

This is a refinement of the simple sifter which permits much higher throughput rates. The vibrating assembly, driven by an adjustable flywheel, is coupled to the suspension to combine high frequency gyration with vertical oscillation. This combined effect gives air suspension to the circulating powder particles.

Fig. 62. The "Vortair" air mounted sieve (Courtesy K.E.K. Ltd.)

Centrifugal Sifting Machines

These are even more efficient than the open sifters for they may be completely enclosed to allow a dust-free and quiet sifting operation.

The material to be sieved is fed into the sifting chamber by a worm discharging the powder into a rotating helical paddle which conveys the pow-

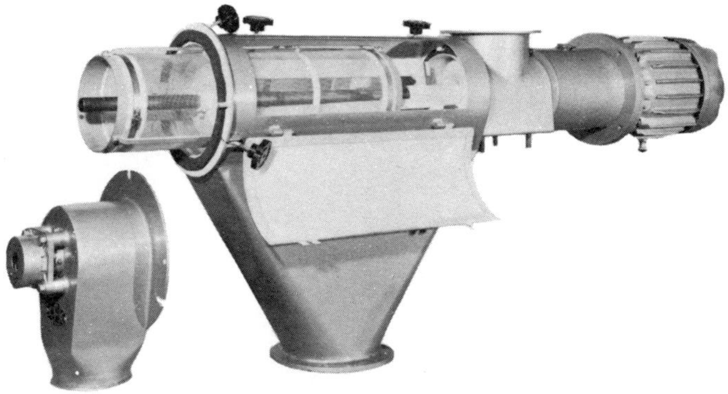

Fig. 63. The S.O. centrifugal sieve (Courtesy K.E.K. Ltd.)

der centrifugally against the cylindrical screen. There is no direct contact between the paddle and the screen so that friction does not wear the screen.

Nylon mesh screens are employed for the sieving of powders with particle sizes in the range 25–250 microns.

Oversize material including fibres and foreign bodies are passed down the length of the cylinder and collected separately.

THE MICRON ULTRA SIFTER

Manufactured by Hosokawa this machine is an efficient cylindrical sieving device with a high throughput at relatively low power ratings.

The cylindrical screen cloth has a surface area of less than 0·3 m^2 with a capacity which is claimed to be 5 to 20 times greater than conventional sifters.

Powder is fed to the cylindrical screen chamber where a high speed multi-blade rotor instantly fluidizes the powder and floats the material to the

Fig. 64. The Micron Ultra sifter (Courtesy Hosokawa Ltd.)

screen by centrifugal force. The housing, screening area and rotor all vibrate in a horizontal and slightly inclined direction to maintain fluidity of the powder. This gentle but effective action permits a high efficiency even with agglomerative powders.

The central parts are held in place by quick release clamps and it takes less than one minute to expose the working parts for cleaning.

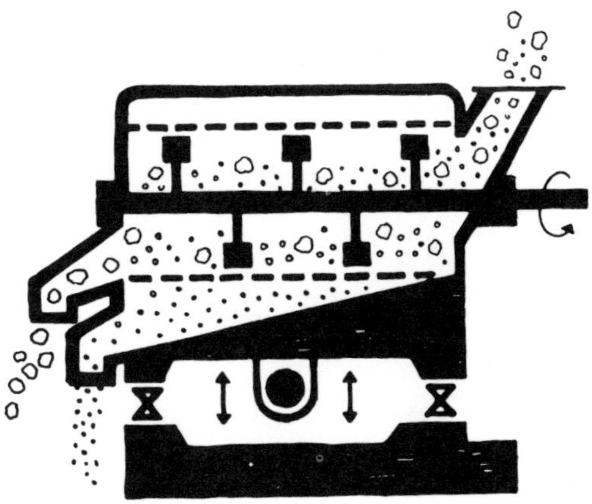

Fig. 65. Operating Principle of the Micron Ultra Sifter (Courtesy Hosokawa Ltd.)

Tumbler Screening Machines

The Allgaier tumbling screens are an improvement upon screen devices previously described since they may be converted from single screens to two or three deck sieves. This makes it possible to sift continuously into four fractions. The cascade arrangement of the screens in a totally enclosed and dust proof assembly allows the coarse particles to be retained by the first sieve and the dust to pass through the final sieve.

Screen mesh sizes are carefully chosen so that the powder particles may be classified between the desired size limits.

Sieves are generally constructed of chromium nickel steel wire and they are cleaned continuously by mechanical or air jet methods to prevent the screens being blocked by fine particles.

Coarser screens are usually cleaned by rotating roller brushes which may be fitted above and below the screen deck. The finest sieve in the bottom deck could easily be damaged by mechanical cleaning methods so that it is common to employ air-jet cleaning methods. This device consists of four nozzle arms rotating beneath the screen. Suction is applied below the bottom screen and air is drawn from the slits in the rotating nozzle arms, first passing upwards through the screen to clear the mesh and then being drawn back through the screen to convey the fine dust to the filter.

The size of sieving machine required will depend upon the particle size range of the classified powder and the throughput rate needed in production.

Fig. 66. Allgaier sieve Model TSM.2000 (Courtesy Alpine (Machinery) Ltd.)

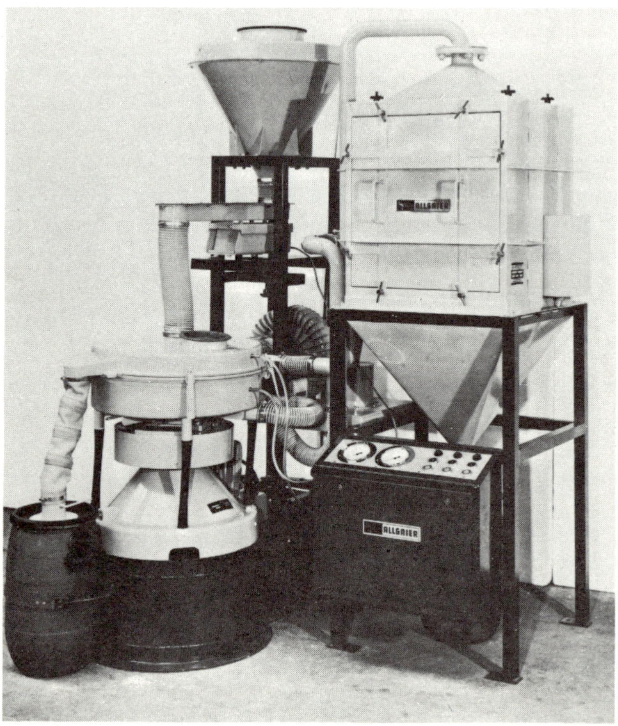

Fig. 67. Allgaier sieve complete with feed and dust collection assembly (Courtesy Alpine (Machinery) Ltd.)

Sieve manufacturers are most helpful in arranging trials to assess each classification problem, and it is best to consult them about this important aspect of powder manufacture.

Centrifugal Air Classification

The range of screening devices so far mentioned are suitable for classification of powders with particle sizes in the neighbourhood of 70 μ but for powders with particle sizes below 50 μ it is most common to employ centrifugal air classifiers.

In these devices particles entrained in a stream of air are caused to enter a spiral vortex flow path created within a stationary walled vessel by the use of guide vanes or tangential inlets. In practice the spiral-vortex air flow path is obtained by using a high speed rotor. Particles carried in the air flow are subjected to two opposing forces, the inwardly directing force derived from the viscous drag of the air, and the outwardly directed centrifugal force derived from their circular motion. Particles in a given size range will be carried in circular paths round and round in the vortex. Those of smaller size will be carried out with the air at the centre while heavier particles will be flung out to the wall and will traverse the walls of the conical container and sink through a bottom exit. The fine particles that are entrained in the stream of air passing upwards through the centre of the vortex pass out of the machine into a filter bed.

This simplified explanation is illustrated by the cyclone (see fig. 68).

The vortical flow is illustrated by fig. 69 and a particle entering this field is acted upon by the centrifugal force proportional to the cube of the particle diameter, and the square of the particle tangential velocity component. It is also affected by the inwardly directing influence due to viscous drag which is more pronounced with the smaller particles.

The tangential velocity component of the particle can be assumed to be equal to the local tangential velocity of the air stream V_t. Velocity of the air stream V may be resolved into radial and tangential components V_R and V_t in accordance with the following

$$V_R = \frac{Q}{2\pi HR}$$

where Q is the rate of volumetric air flow; H is the rotor width and R is the radial position.

At the same point R the tangential velocity can be expressed by

$$V_t = KR^n$$

where K and n are constants.

The viscous drag effect is directly proportional to the diameter of the particle and the relative velocity of the air and the particle in the radial direction.

There will obviously be a point at which a critical particle size will be affected equally by the two opposing forces and this is described as the cut size.

This cut size may be calculated from the equation.[1]

$$D_{50\%} = \frac{3v}{\pi D_0 \omega}\sqrt{\frac{\eta Q}{\pi H \rho_p}}$$

where D_0 is the rotor diameter, η is the viscosity of the air stream and ρ_p is the true density of the particle ω is the rotational speed of the rotor. v is a

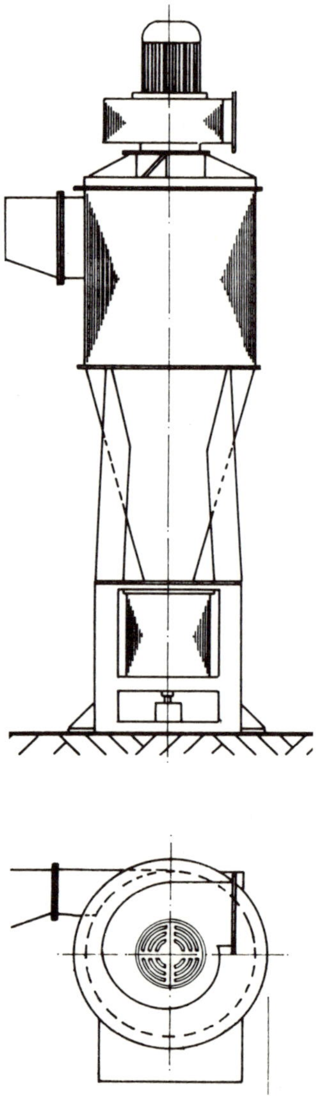

Fig. 68. Cyclone collector (Courtesy Drury Powder Ltd.)

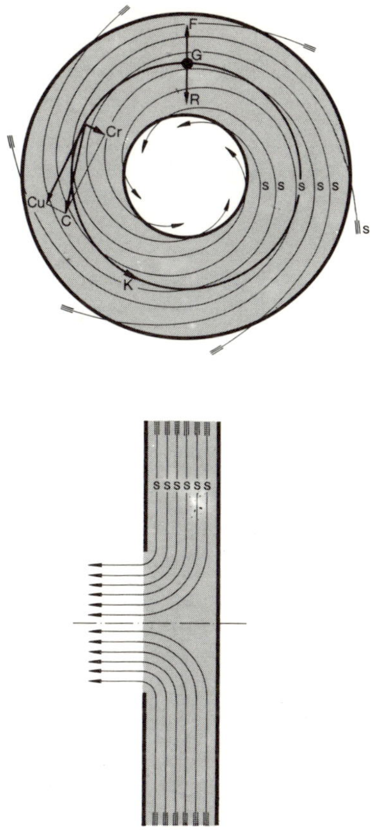

Forces in the spiral flow
F = centrifugal force, R = frictional force, G = cut size particle, K = circular path, s = spiral flow line, C = air velocity, C_u = peripheral component, C_r = radial component.

Fig. 69. Idealized centrifugal flow (Courtesy Alpine (Machinery) Ltd.)

constant derived from the flow pattern for the rotor which must be determined experimentally.

The important conclusions which may be drawn from the above equation are:

(a) Particles of low density have an increased cut size so that it will be more difficult to remove the fines from low pigmented powder coatings.
(b) An increase in the flow rate of the dust laden air increases the cut point and
(c) Increasing the rotor speed does decrease the cut size.

There are several commercially available air classifiers and a description of one of these machines illustrates the principles of this technique.

THE ALPINE MIKROPLEX SPIRAL AIR CLASSIFIER

This is a compact machine with a vertical classifying chamber. Particles entering the classifier (1) fall on to the feeder plate (2) where they are drawn at speed and deflected by the bumper plate (3) into the classifying chamber (4). In this zone a radial flow fan (5) causes separation of the particles with the larger sizes thrown to the outside by centrifugal force, where they fall into the coarse particle chamber (15) by deflection at the coarse product brake ring (13). The air stream is forced through the rejecting blades (14) into the classifying chamber where it is drawn through the upper opening of the cross flow fan (6) into the cyclone (8). Air freed from dust is then recirculated into the classification chamber.

The cut point depends primarily upon the rotational speed of the cross flow fan, the adjustment of the brake ring and the feed rate.

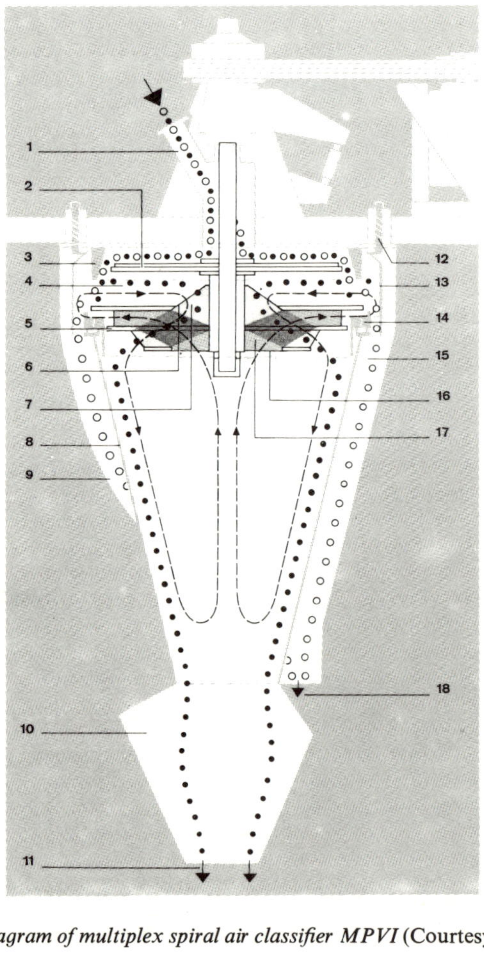

Fig. 70. Schematic diagram of multiplex spiral air classifier MPVI (Courtesy Alpine (Machinery) Ltd.)

Fig. 71. Cross flow fan with feeder plate (Courtesy Alpine (Machinery) Ltd.)

Fine particles removed by classification methods may be incorporated in the premix of the following batch and recycled.

MICRON SEPARATOR N TYPE

An efficient alternative air classifier is the Hosokawa Micron Separator N Type which consists of an upper classifying chamber, an elutriation chamber and a fluidized bed.

Principle of Operation

The material fed into the Separator is first dispersed in the fluidized bed and conveyed to the classifying chamber where classification takes place at a preselected cut-point. The fines are conveyed to a suitable collector through the upper outlet while the coarse particles descend along the inner wall of the housing. In the course of their descent the particles pass into the elutriation chamber where they are subjected to violent turbulence. Agglomerates of

Fig. 72 Brake ring with guide vanes (Courtesy Alpine (Machinery) Ltd.)

Fig. 73.

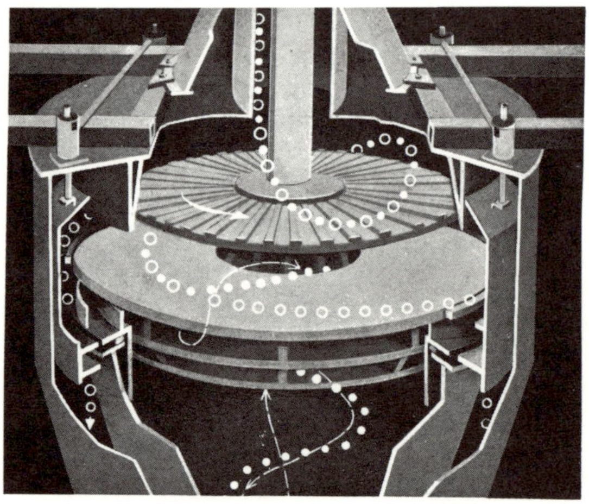

sectional drawing – Mikroplex spiral air classifier type MPVI

o o o coarse product

• • • fines

← – – classifying air

Fig. 74. Sectional Drawing Mikroplex MPVI (Courtesy Alpine (Machinery) Ltd.)

Fig. 75. Micron separator N Type (Courtesy Hosokawa Ltd.)

particles are broken up and the debris lifted back into the classification chamber for further classification. The coarse particles after passing the elutriation chamber are discharged through a rotary air lock valve.

The model most suitable for handling the volumes produced in powder coatings manufacture has the following specification.

Table 15. Specification of the Micron Separator N Type
(Courtesy Hosokawa Ltd.)

Model	Overall Dimensions		Power	Maximum Rotor Speed	Air Volume	Input Capacity
	Diameter	Height				
MS 1N	350 mm	1,500 mm	0·75–1·5 kW	2,300	10 to 15m^3 per min.	1 to 1·5 tonnes/hour

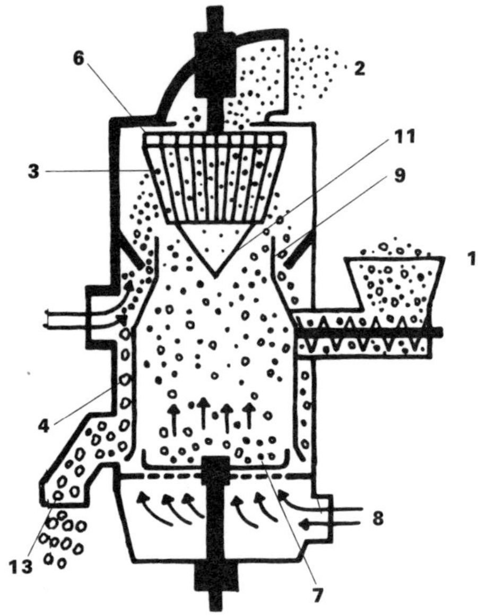

Fig. 76. Operating principle of the Micron separator N Type (Courtesy Hosokawa Ltd.)

In-line Classifiers

In Chapter 7 an account was given of the screenless fine grinding machines having built-in classifiers to prevent coarse particles leaving the grinding chamber. While this system is effective in preventing coarse particles being carried over to collection, it does tend to produce a higher proportion of fines, approximately 5% below 15 microns. This quantity of fines does not adversely affect the performance of powder coatings applied by hand guns, but it can cause blocking of feed lines and injectors with continuously operating automatic powder coating units.

The removal of this small amount of "fines" can be achieved by installing a separator in the powder feed line between the mill and the collection chamber.

Suitably classified powder with the coarse and fine particles removed can be collected and packed at the "tailings" discharge point of the in-line classifier. The small amount of fines can be left in the classifier and, to avoid the time consuming process of cleaning the collection unit between batches, the fines of all colours can be mixed in the collector and used later in a premix for a neutral grey powder.

THE MICRON SEPARATOR

Hosokawa are the manufacturers of the Micron Separator which offers high capacity and high efficiency at cut-points ranging from 140 microns down to 5 microns.

It is a compact machine which is easy to operate, clean and maintain.

Principle of Operation

Material enters the Micron Separator through the intake (1) and, rising through the adjustable inner feed pipe (9), flows over the rotor guide cone (11) to the classifying rotor (3) which is motor driven at variable speeds. This variable speed adjustment enables a wide range of cut points to be obtained. Particles of the required fineness are carried through the rotor airway slits (6) and conveyed to a suitable collecting apparatus through the outlet (2).

Oversize rejected particles descend and are rejected through the "tailings" duct (4). During their descent they are subjected to a violent turbulence caused by the elutriation ring (7) and the incoming secondary air entering the Separator through duct (8). This washing action causes a secondary refining which removes the fines agglomerated with the larger particles.

The tailings discharge (13) may be equipped with a rotary valve.

Table 16. Range of Micron Separators
(Courtesy Hosokawa Ltd.)

No.	Power (h.p.)	Air volume (m^3/min)	Capacity (kg/hr.)
MS-0	½	2–6	10–15 m^3
MS-1	1	6–15	50–150
MS-2	2	20–50	150–500
MS-3	3	60–100	500–1,000
MS-4	5–7½	100–200	1,000–2,000
MS-5	7½–10	200–300	2,000–5,000

It is most important to ensure that the collection unit is capable of handling the high volumes of air required by the selected Micron Separator. Generally, it is not possible simply to insert a Micron Separator into an existing production line comprising a screenless mill and a Pulsaire collector.

Dust collection units are described in Chapter 11 and these may provide a cheaper and more effective collection point for the small amount of fines exhausted from the Micron Separator.

THE 'MICROSPLIT' CLASSIFIER

"Microsplit" separator units are manufactured by British Rema for the classification of powders which fall within the particle size range 3 to 75 microns. These units offer a high output rate relative to the small size of the separator, with high extraction efficiency and low power consumption. It is a completely self-contained unit carrying its own circulating fan, separating element and coarse and fine collecting cyclones. There is no need to purchase additional cyclones or filter units and the separator can be operated in closed circuit with most types of fine grinding machinery where the product is discharged continuously from the rotary valve and may be screw fed directly to the "Microsplit" separator.

The operating principle of the "Microsplit" is shown in fig. 77. Material to be classified enters the separator down the central vertical feed pipe (1) and is evenly dispersed by the spinner disc (2) into a rising air stream which carries the particles up to the multivaned separating rotor (3). As they approach the rotor, the particles are subjected to high centrifugal forces which compete with the centripetal frictional drag forces exerted on particles by the circulating air as it spirals inwards through the rotor vanes. For the particle sizes represented by the cut point, these two opposing forces are

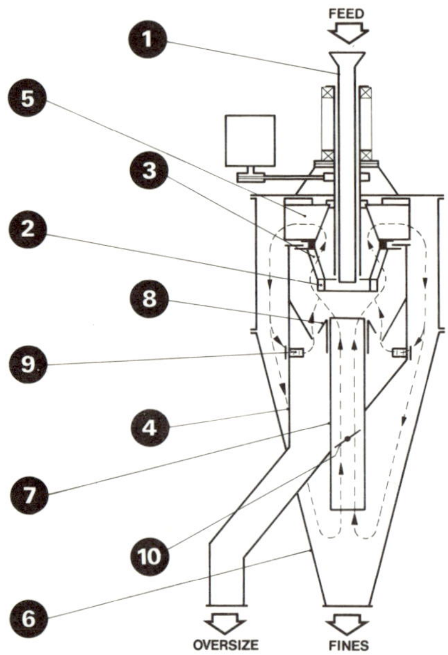

Fig. 77. Microsplit operating principle (Courtesy British Rema Ltd.)

balanced; the centrifugal force dominates, however, on particles above the cut size causing them to fly away from the rotor, to impinge on the inner casing and descend to be collected in the oversize cyclone (4).

Particles below the cut size are conveyed inwards, by the influence of the centripetal drag force (dominant on small particles), through the circulating fan (5) to be separated from the air stream in the fines cyclone (6) of the outer casing. The cleansed circulating air returns to the separating rotor via the central return pipe (7).

The efficiency of separation of any classifier depends largely on its ability completely to disperse each individual particle into the separating zone and with most materials this becomes increasingly difficult as the cut size is reduced. Fine particles tend to agglomerate and be classified as particles of the agglomerate size. To overcome this tendency a secondary air system is built into the "Microsplit" separator. The secondary air system consists of a series of adjustable vents (9) built into the inner casing which allow a percentage of the circulating air to return to the separating region via the rejects cone. This air sifts the rejected material as it descends to the rejects cone liberating the fine particles and returning them to the separating rotor. By adjusting the area of the vents it is possible to control the volume of secondary air to the optimum for any desired cut point of feed material. The effect of the secondary air system is further enhanced by an adjustable sleeve (8). This controls

Fig. 78. Microsplit separator Size 3 (Courtesy British Rema Ltd.)

the area of the annular gap between the air return pipe and the inner cylinder, and hence the air velocity at that point. Variation of the cut size is achieved by adjustment of the circulating air or by changing symmetrically the number of blades on the separating rotor.

In general, maximum separating precision is achieved with a full complement of blades fitted but, where product requirements are less critical, the number may be reduced to give increased output.

Both the primary air, which returns to the rotor via the control air pipe containing butterfly valve (10) and the secondary air, are easily controlled by external hand wheels which operate against graduated scales.

THE "AEROSPLIT" CLASSIFIER

This machine is a direct development of the "Microsplit" separator which is designed to operate either in open circuit as a packaged classifying unit complete with its own filter and fan system, or in closed circuit systems.

Generally, the "Aerosplit" classifier is used to separate particles which are being pneumatically transported from the fine grinding machine and the principle is illustrated in fig. 80.

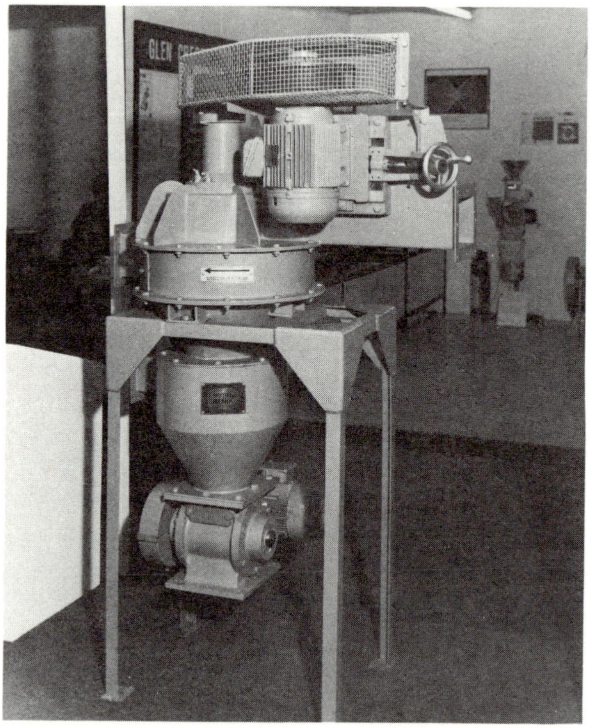

Fig. 79. Aerosplit classifier Size 1 (Courtesy British Rema Ltd.)

A secondary air system is also included in the "Aerosplit" classifier.

Control of product fines with the "Aerosplit" classifier is achieved by varying the speed of the separating rotor. The sharpness of separation and efficiency of this machine are exceptional and where the oversize from the classifier is the desired product, this can be produced to contain a minimum of dust.

Since the system operates under vacuum conditions a very clean plant can be engineered.

Both the "Microsplit" and the "Aerosplit" are forced vortex machines and as a result the fineness of the fines product remains constant over large feed rate variations.

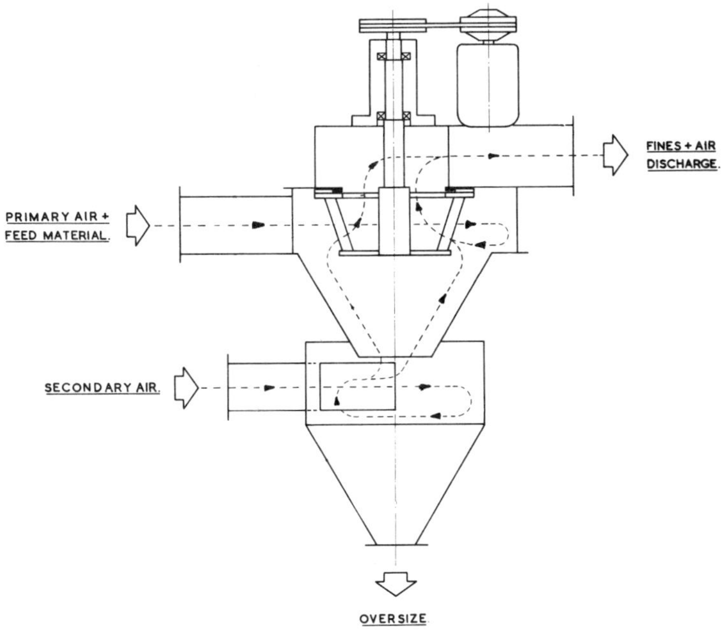

Fig. 80. Operating principle of Aerosplit classifier (Courtesy British Rema Ltd.)

SIZE ANALYSIS OF FINE PARTICLES

A considerable number of methods have been devised for measuring the size of fine particles. These include sieving, sedimentation and elutriation, microscopic analysis, light scattering, gas adsorption, impact and sensing zone methods.

Particle size analysis is often a time consuming process although modern methods with up-to-date instruments have greatly improved the assessment by microscopic sedimentation and sensing-zone techniques.

Particle size distribution may be expressed by diameter, area, volume, weight or by number and the manner in which each of these is defined can affect the efficiency of the analysis. The excellent paper presented by Ullrich[(2)] illustrates the problems of assessing the sizes of particles in the range 0·5 to 300 microns, where the ratio of diameters is 600:1, the ratio of areas is 360,000:1 and ratio of volumes is 216,000,000:1. A further complication is the particle shape. Obviously it is difficult to define a needle shaped particle by particle diameter.

Those analytical methods which require the particle to be suspended in a liquid must rely upon efficient dispersion techniques to ensure that flotation and flocculation is minimized, that agglomerates are not present, and the particles must not be soluble in the liquid.

The sample chosen for particle size assessment must also be representative of the bulk material, and this is more difficult to achieve when a wide range of particle sizes exists in the powder sample.

Results of particle size analysis must be presented in an acceptable and usable form and there are more than 100 ways of expressing the analysis in a graphical manner. Plotting methods have been discussed in detail by Herdan and Smith.[3]

The following methods are briefly described:

Microscropic Analysis

Sophisticated techniques for the assessment of the size distribution of ultra-fine particles are available but it is sufficient to consider the standard microscope technique, or better still, that using the ultramicroscope for resolving individual particles in the range 0·25–250 microns. Powder spread out in a single layer on a glass slide is compared under suitable magnification with reference circles of known size inscribed on a graticule which may be viewed at the same time as the particles. A number count of particles in each size range is made and the accuracy depends upon the number of separate assessments. The numbers are best expressed as percentages in terms of cross sectional area. The method is laborious.

Fisher Subsieve Sizer

This instrument is a self calculating air permeability apparatus which assumes that the surface area is due to spherical particles so that the mean diameter may be evaluated without mathematical computation.

Air at a constant pressure is passed through a compacted bed of powder and then through two flow meter resistances. The pressure drop across the resistances is indicated by a water manometer and the combination constitutes a flow meter. Simultaneously, the difference between this pressure drop and the initial applied pressure represents the pressure difference across the bed of powder.

The specific surface area (in $cm^3\ ml^{-1}$) is given by.

$$S = k\frac{(AL-1)^{\frac{3}{2}}}{L}\left(\frac{P-F}{P}\right)^{\frac{1}{2}}$$

where k is a constant,

A is the cross sectional area of the bed in square cm,

L the height of powder in cm,

P the initial pressure applied to the powder bed in cm of water,

F is the pressure difference across the flow meter resistance.

The specific surface area of a powder may be equated to $(6 \,.\, 10^4)/d_m$ where d_m is the mean diameter in microns.

$$d_m = K\,.\,\frac{L}{(AL-1)^{\frac{3}{2}}}\left(\frac{P}{P-F}\right)^{\frac{1}{2}}$$

The instrument is supplied with a calibrator which enables the average particle size to be read directly from the calibrator chart.

An average particle size in the range 0·2 to 50 microns may be quickly obtained by this method.

Sedimentation Method

This hydrodynamic method is dependent upon Stokes Law for the calculation of particle size in the 0·5 to 300 microns range.

$$d = \left[\frac{18\eta h}{(\sigma - \rho)gt}\right]$$

where η is the absolute viscosity of the medium in poises, ρ is the density of the medium in g/cm^3

σ is the apparent particle density in g/cm^3,
h is the distance particle falls in time t (sec),
g is the acceleration due to gravity (cm/sec^2).

Continuous readings are taken by a sedimentation balance. This method is effective in the medium particle size range (where the rate of sedimentation is high) but becomes slow and tedious with superfine particles.

A similar method which allows sedimentation of particles from a gaseous medium is the aerodynamic process of elutriation. This permits automatic recording and is generally faster than the hydrodynamic method.

Air Jet Sieving Methods

Sieving is a simple and rapid method of size analysis giving results directly in terms of the weight distribution of sizes. The linear size assigned to a particle is the length of the side of the minimum square through which the particle will pass. For irregular particles, not too extreme in shape, the numerical value of the sieve diameter is within about 10% of the volume diameter and the Stokes diameter, but is only about 0·7 times the projected diameter of settled particles.

The efficiency of the sieving action has dictated the lower size limit since it must overcome particle adhesion, and the finest mesh sizes have been difficult to produce uniformly. Mechanical sieving has been in general restricted to measuring particle diameters of 75 microns and above.

Two recent developments have enabled particle diameters as small as 20 microns to be measured: these are the introduction of the airjet sieving method and the manufacture of really fine mesh by electro-deposition processes.

The technique of airjet sieving has been described on p. 129 and a suitable machine is supplied by Alpine (Machinery) Ltd.

Generally speaking the analysis is simple and rapid for particle diameters in the range 32 to 75 microns, but can become very slow for particles whose diameters are approaching 20 microns.

Sensing Zone Method by Electrical Resistivity

This is an account of the basic principle of the Coulter Counter which is a rapid method of particle size analysis in the superfine and ultrafine range. In a few minutes it can completely analyse a sample containing particles whose sizes range from 1 to 75 microns.

The particles are suspended in an electrolyte solution in which they are insoluble, and drawn through a small aperture in a non-conducting wall. The wall has an immersed electrode on each side and a current flows from one electrode through the aperture to the other electrode. As each particle passes through the aperture it displaces its own volume of electrolyte and so modulates the electric current passing through the aperture. This modulation can be detected as a change in voltage, current, resistance or impedance, the magnitude of the modulation being proportional to the volume of the particle. Typically in modern commercial equipment, current or voltage changes are measured.

As up to 5000 particles pass rapidly through the aperture every second, the particles produce a series of electronic pulses which can be scaled and

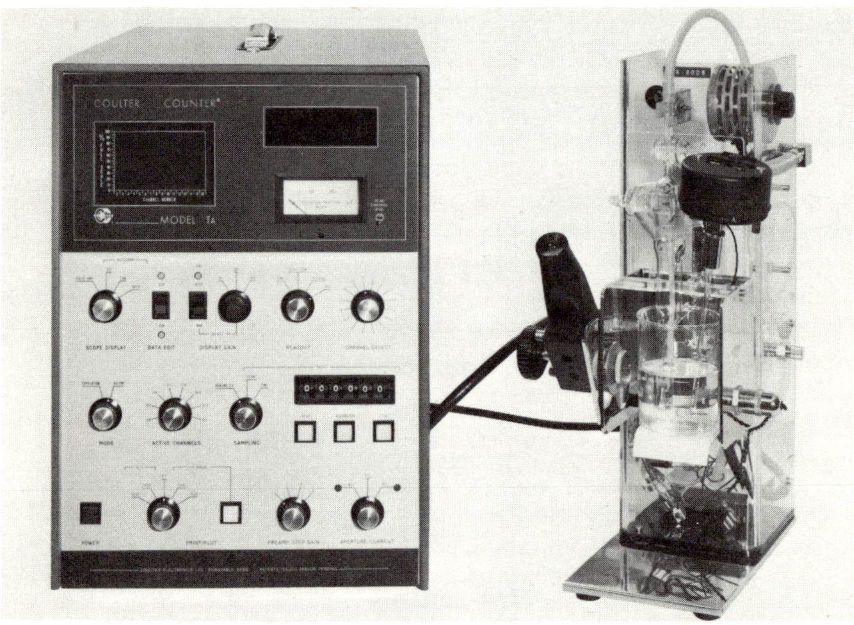

Fig. 81. Coulter counter Model TA (Courtesy Coulter Electronics Ltd.)

counted, to give (volume) size and number information. The various models of Coulter Counter all perform this basic function, but with varying degrees of sophistication.

Any aperture will sense and be used to measure particles from 2% to 40% of its own diameter, except when the particles have extreme shapes, when the range will be somewhat reduced. Powder coatings normally are in the size range 3 to 70 microns and a 140 or 200 micron aperture would be best suited.

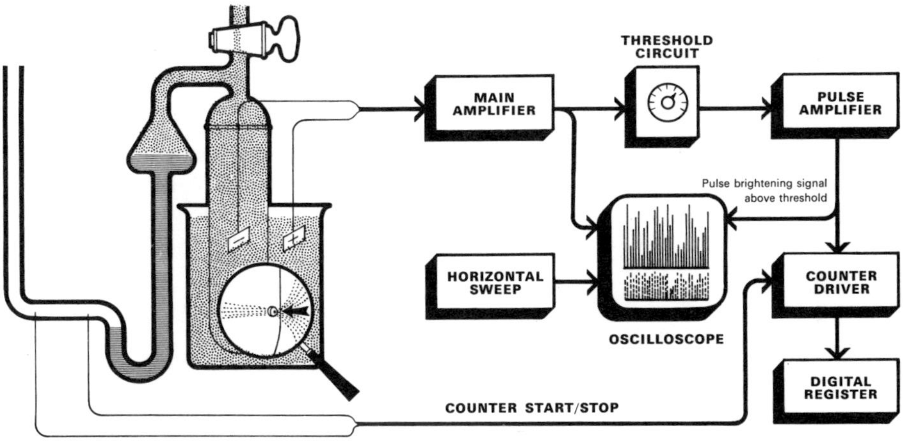

Fig. 82. Principle of Coulter Counter process (Courtesy Coulter Electronics Ltd.)

A particularly fine powder, of 1 to 20 microns, could be analysed using a 50 micron aperture. A wide range of particle sizes from 1 to 75 microns can be readily measured using two apertures which are easily interchangeable.

The model TA is the most advanced industrial model to date. It is a multi-channel device, so that only one pass of sample suspension through the aperture is needed to provide up to 16 channels of simultaneous data each channel displaying the number of particles in a given size range passing through the aperture. An oscilloscope presents the size distribution as a cumulative or differential weight (volume) histogram simultaneously with the analysis itself.

It is essential to ensure an adequate dispersion of the material under test in the suspending electrolyte. Small ultrasonic bench units are able to produce an adequate dispersion in 15–20 second without any risk of attrition which would occur if mechanical methods were used. Recognized dispersants such as Calgon and other sodium phosphates can be used as the electrolytes.

The response of the Coulter Counter is relatively independent of particle shape over the size range covered by each orifice tube, since the method determines the true volume of each particle as it passes. It has been shown that the change in aperture resistance caused by a particle is

$$\Delta R = \frac{P_0 V}{A^2}\left\{\frac{1}{\dfrac{1}{1-\dfrac{P_0}{P}} - \dfrac{a}{xA}}\right\}$$

where P_0 is the electrical resistivity,

A is the aperture area normal to the axis,

V, P and a are particle volume, effective resistivity and area normal to the aperture axis,

x is the particle dimension ratio which is given by

$$x = \frac{\text{the length along the aperture axis}}{\text{diameter of an equivalent sphere}}.$$

The errors are not great, being of the order of one per cent on the volume.

A sample weight of 10–80 mg of powder per 100 ml of electrolyte is suitable for most assessments.

COULTER COUNTER

Particle Size Analysis Results

MAT E100M Batch 3 (Normal Grade) SOURCE DATE 12·8·75

ELEC. ISOTON ORIFICE 280μ DISP. Nonidet P42 ultrasonics

INSTRUMENT CONTROL SETTINGS		EQUIVALENT PARTICLE DIAMETER	NO. OF PARTICLES ABOVE SIZE 'D'	CUMULATIVE % wt. ABOVE 'D'
A	W	Dμm	N	
354	0	102	0.5	0.41
177	0	80.6	5	2.03
	1	64.0	33	7.73
	2	50.8	156	19.9
	3	40.3	505	37.8
	4	32.0	1159	54.5
	5	25.4	2199	67.5
	6	20.2	3738	77.2
	7	16.0	5977	84.6
	8	12.7	9114	89.4
	9	10.1	13913	93.1
	10	8.0	20811	95.9
	11	6.35	30326	98.0
	12	5.04	43049	99.2
	13	4.00	53513	99.6

Fig. 83. Typical powder coating analysis (Courtesy Coulter Electronics Ltd.)

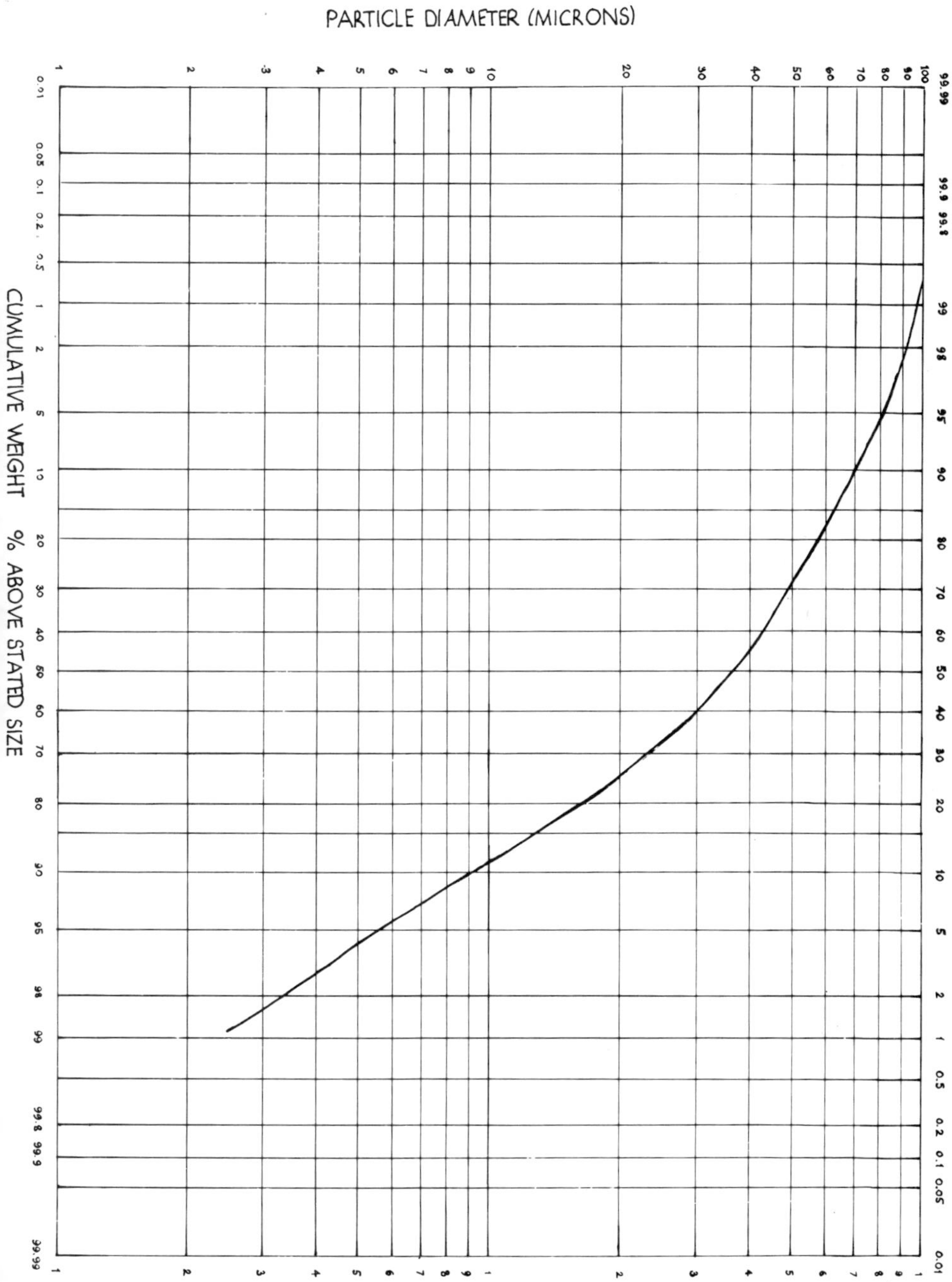
PARTICLE DIAMETER (MICRONS)
CUMULATIVE WEIGHT % ABOVE STATED SIZE

PARTICLE SHAPE

The ideal shape of a powder coating particle is spherical but it is difficult to achieve this shape by the present methods of comminution because of the brittle nature of the low molecular weight powder coating resins.

Spherical particles may be obtained when the powder is produced by spray drying techniques but it will be difficult to produce a more uniformly spherical particle by the conventional methods of powder manufacture unless the base resins are increased in molecular weight thereby making them less friable. The replacement of direct attrition grinding by shear methods in the presence of a liquid carrier will also assist the development of uniformly spherical particle shape.

This is an area of development which must be investigated if the thin film powder coating film is to become a reality.

COHESIVE FORCES

These inter-particle forces mainly arise as the result of Van der Waals interactions and they result in agglomeration and reduced fluidity of the particles.

Cohesion increases as the particle size decreases and it will vary considerable with relative humidity. Electrostatic charges, even if these are dissipated quickly, may bring particles within the range of Van der Waals forces. A simple screening of the powder before application, by the use of a vibratory coarse screen, should be sufficient to annul temporarily the effects of cohesive forces.

It is often stated that the presence of moisture in a powder coating is harmful but it can be shown that a moisture content as high as 1% does not adversely affect the flow characteristics of most powder coatings. In fact, a small quantity of water (in the region of 0·4%) can be most beneficial.

Whilst it has been stated that relative humidity can promote cohesive forces, this is only a minor contribution to poor flow properties. Thin layers of moisture adsorbed on the particle surface can promote free flow by imparting a lubricating effect, although it is equally true that thick films of moisture can bind particles together by surface tension.

The presence of moisture at the correct level will enhance the charge acceptance qualities of a powder and assist the eventual, but necessary, dissipation of charge. A completely dry powder containing a small quantity of silica gel will show poor charge acceptance properties.

The essential presence of water is one of the reasons why powder sprayers are instructed to mix oversprayed powder with fresh powder, before using reclaimed powder.

Small amounts (up to 0·5% by total weight) of finely divided silicaceous materials can improve the flow properties of powder coatings which normally agglomerate during manufacture.

Where an excessive layer of moisture causes the particles to be bound together by surface tension, the presence of these finely divided materials may assist the flow characteristics by disrupting the cohesive film. These flow aids are generally incorporated by dry blending techniques and for this reason it is advisable to blend new and reclaimed powder.

POWDER FLOW AND THE METHODS OF ASSESSMENT

If a tightly packed powder is caused to flow, it can be shown that some degree of dilatency is a necessary first step in order that the mass of particles change shape and begin to move.

The successive steps for the flow of powder from an unrestricted container have been detailed by Jenike.[4]

Powder coatings with poor flow characteristics within the container will exhibit the effect shown in fig. 84(g) which is commonly referred to as "rat-holing".

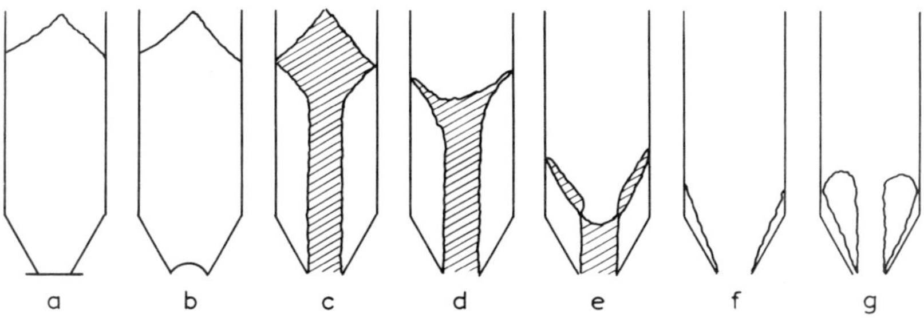

Fig. 84. Flow from an unrestricted container

It is commonly accepted that the angle of repose of a heaped powder is a direct measure of its ability to flow and powders with angles of repose less than 40° are said to flow easily.

Cohesive powders of less than 100 micron diameter have high angles of repose, but the test it easy to apply and may be used as a simple method for the assessment of batch to batch reproducibility.

The angle of repose can vary considerably according to the way in which the powder is poured, and it is best to assess this property by a drainage method.

A flat, circular, steel plate with a smooth surface of known diameter, is mounted upon a tripod in the middle of a wooden platform which has a central drainage hole. The tripod legs are adjusted so that the surface of the plate is level, and the hole in the platform is closed by a cork. A large tin-plate open-ended cylinder is placed on the platform to surround completely the cylindrical plate and tripod (fig. 85).

Powder is carefully poured into this cylinder and it is then allowed to drain through the hole in the wooden platform. The height of the apex of the powder remaining on the cylindrical plate is measured by means of a pointer attached to a retort stand.

$$\text{The angle of repose } \alpha = \tan^{-1}[2h/D]$$

where h is the axial height of the cone of powder,
D is the diameter of the cylindrical plate.

Wide variations in the angle of repose may be due to compaction of the powder or to a high moisture content.

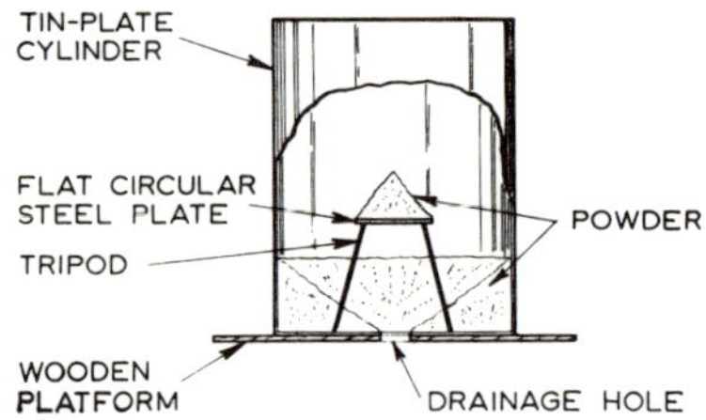

Fig. 85. Device for measuring the drained angle of repose

Experiments have been made with a series of open-ended glass cylinders with different outlet aperture sizes, but the results have been too inconsistent for meaningful comparison.

A completely practical method for the assessment of powder flow is to place the powder into the feed hopper of the electrostatic hand gun, then disconnect the powder feed line from the gun and allow the powder to be transported from the base of the hopper and returned to the top of the hopper. If there is no surging or blocking of the powder during a ten minute cycle it is unlikely that difficulties will be encountered during application.

There may be some merit in the development of a testing apparatus employing the principle of the Jenike shear cell (fig. 86) which would enable the angle of internal friction of the powder to be determined. This is a more relevant measurement of cohesive powders than the assessment of angle of repose.

The test equipment consists of a shallow cylindrical cell which can be split horizontally.

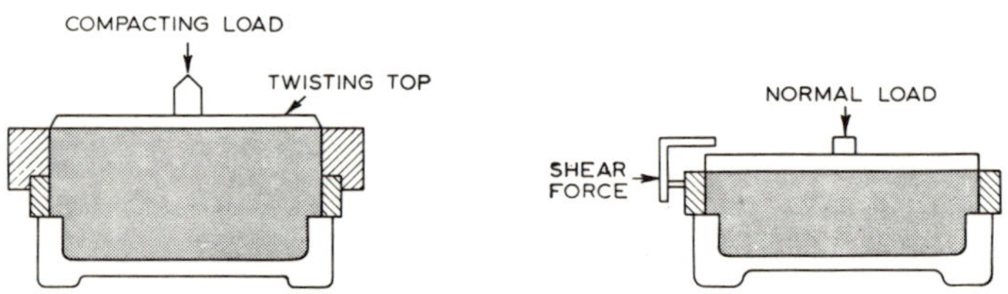

Fig. 86. The Jenike cell

A sample of powder to be tested is first compacted by the application of a compacting force to the lid of the cell. This compacting force must be carefully chosen to avoid the limiting degree of compaction beyond which the side shear cannot be applied. Compaction is necessary to ensure that the volume does not change during the test. The lower half of the cell is fixed and a shear force is applied at a constant low rate to the upper section which is free to move. A number of different loads may be applied to give a relationship between normal stress and shear stress. These may be plotted to give a yield locus (fig. 87) where δ is the effective angle of friction.

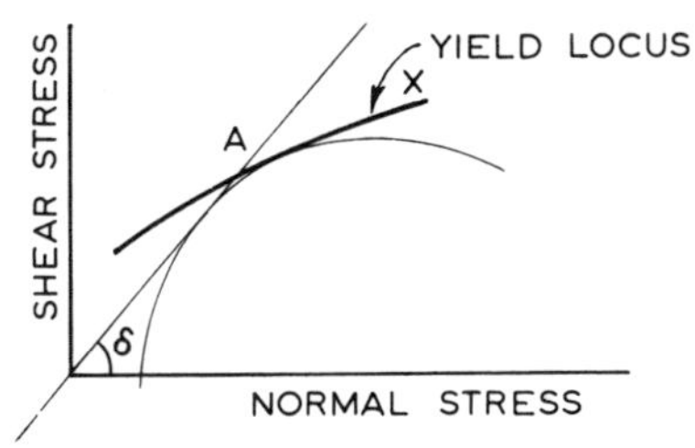

Fig. 87. Graph of yield locus

There are several interesting observations to be drawn from this graph.

1. The yield locus does not extend indefinitely, but terminates when the ratio of the normal stress to shear stress reaches a critical value beyond which compaction occurs.
2. The method by which the powder is consolidated will define the position of the yield locus.
3. Extrapolation of the yield locus to the shear axis gives a measure of cohesive strength.
4. Extrapolation of the yield locus to the normal stress axis gives a measure of the tensile strength of the powder.

It is also possible by this method to calculate the angle of friction between the powder and the wall of the container thereby giving a measure of the adhesive forces.

REFERENCES

1. H. W. Newton and W. H. Newton. "A Study of Classification Calculation", *Rock Products* **35** (1932).
2. O. A. Ullrich. Paper presented to Instrument Society of America Conference, New York (1960).
3. G. Herdan and M. L. Smith. *Small Particle Statistics*, Elsevier Publishing Company, New York (1953).
4. A. W. Jenike. Bulletin 108 of the Engineering Experimental Station, University of Utah (1961).

9

Powder Coatings Applied by Fluidized Bed

SUMMARY

The phenomenon of fluidization of powders has been extensively researched over the last fifty years so that the problems of handling and conveying finely divided powders could be fully understood, and industrial methods developed.

Fluidization is a term used to describe a process in which a rapid gas-current passes through a finely powdered substance in such a manner that the powder particles become separated and airborne without being carried away by the air stream. The fluidized powder bed then behaves like a fluid.

It was not until 1952 that research workers at the German firm Knapsack-Griesheim developed this principle as a method of dipcoating metallic articles with thermoplastic powders.[1] Within ten years this process had become widely accepted as an industrial coating operation and extended to include the fluidized bed application of thermosetting powder coatings.[2]

Thermoplastic powder coatings, however, still dominate this field of application and the commercially acceptable coatings include nylons, epoxies, vinyls, polyethylene, chlorinated polyethers and cellulosic compounds.

It has been mentioned in a previous chapter that the high molecular weight thermoplastic resins are characterized by excellent toughness and chemical resistance, although the adhesion of their films is generally inferior to the adhesive qualities shown by the thermosetting resins. The adhesion of thermoplastic films may be considerably improved by the application of a primer so that remarkably tough, chemical resistant coatings with good adhesion are now available.

Articles to be coated by this dip method are heated to a temperature exceeding the softening point of the powder coating, the particles melt, and adhere to the surface of the articles when they are immersed in the fluidized bed.

An alternative method, which avoids the necessity to preheat, is the electrostatic fluidized bed whereby the powder particles obtain an electrostatic charge and these are attracted to the article passing over the fluidized bed since it provides a site of opposite polarity.

Only articles of relatively simple shape have been successfully coated by the electrostatic fluidized bed and the unmodified fluidized bed is still the predominant process.

The particle size of powders suitable for fluidized-bed application lies in the range eighty to two hundred microns.

Film thickness will depend largely upon the particle size of the powder, the temperature of the article and the time the article is immersed in the fluidized bed. Generally, a coating thickness of three hundred to four hundred microns is obtained.

The appearance of the coated surface is affected by the manner in which the coated article is cooled. A satin finish is often obtained by air cooling, whilst water quenching results in a high gloss finish.

It is a golden rule that successful coatings may only be applied to a correctly cleaned surface, and dirt or grease must never be encapsulated by fluidized bed-applied coatings.

THE PRINCIPLES OF FLUIDIZATION

The basic principles of fluidization are deceptively simple, for they merely entail partially filling a vessel with a porous bottom (referred to as a "bed") with powdered plastic material, and then maintaining this material in a fluidized state by an adequate flow of gas, usually air, through the porous base plate. In the fluidized condition the powder shows many of the properties of a fluid and one of its most important characteristics is that of particle mobility. The bulk density of the powder is reduced in the fluidizing bed so that the particles can move freely within the bed.

It is essential for the correct application of coatings that the bed operates smoothly without any settling or surging of powder.

The stable state in which the pressure drop across the bed is just sufficient to support the weight of the particles is referred to as incipient fluidization.

Figure 88 shows the linear increase of pressure with increasing air flow before the steady state of fluidization is achieved when the pressure drop becomes constant. This settled state of the particles is defined as a fixed bed. It will be noticed that the point of incipient fluidization is followed by the attainment of a lower pressure drop (which remains constant with increasing

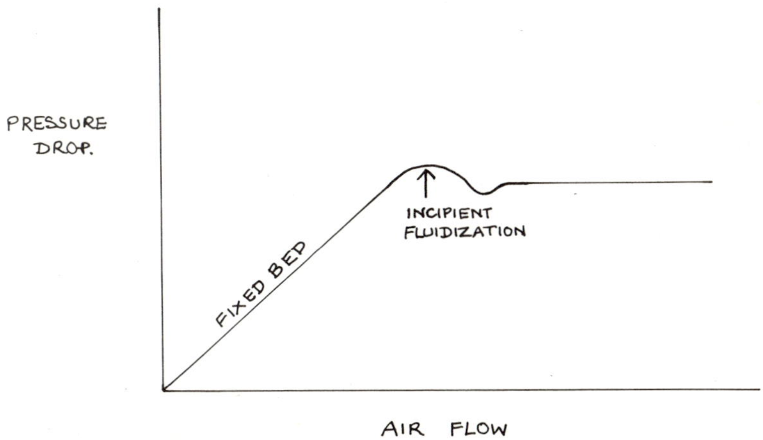

Fig. 88. Diagram illustrating the attainment of incipient fluidization

air flow) and it is possible that the slight reduction in pressure drop following fluidization is connected with the dilatency shown when solid particles begin to exhibit fluid flow.

Channelling is a term used to describe uneven air flow through the fluidized bed which is often caused by the presence of very fine particles or cohesion between the particles resulting in agglomeration. This defect can be reduced by vibrating the fluidized bed or by the introduction of some form of mechanical agitation.

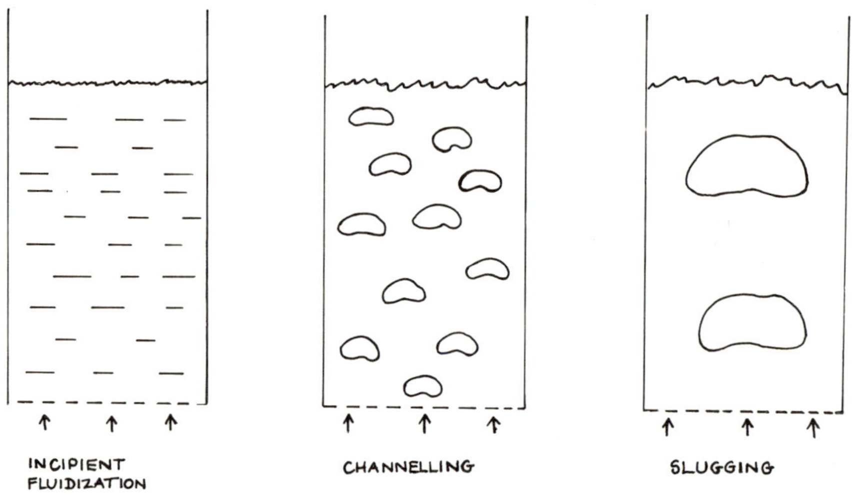

Fig. 89. Diagram—incipient fluidization, slugging and channelling

Slugging or spouting is caused by the surge of a large gas bubble through the bed which carries away powder from the fluidized container. This problem is often resolved by increasing the diameter of the fluidized bed.

The voidage fraction is the proportion of the bed volume that is made up by the fluidizing gas. When the voidage fraction is greater than 0·8 the powder will be conveyed out of the bed by pneumatic transportation. For prolonged smooth operation, it is essential that the gas used is dry and free from oil contamination.

THE FLUIDIZED BED DIP COATING PROCESS

At the bottom of the fluidizing tank there is a plenum chamber separated from the tank proper by a porous tile wall. The plenum chamber, is fed with low pressure air and the powder is fluidized by the air diffusing uniformly up through the porous wall; which is made of sintered bronze, 2 to 5 mm thick with an average pore size of 12 microns. This is the preferred construction material.

The porous tiles are supplied in standard dimensions and larger tanks may be constructed by joining together several standard tiles with soft solder.

Air is provided in the perfectly clean condition required for smooth operation of the bed by a grease-free dry-segment compressor. It is extremely difficult to remove traces of oil from a normal compressor even when used in conjunction with a sedimentation tank or dessicator unit. The blower

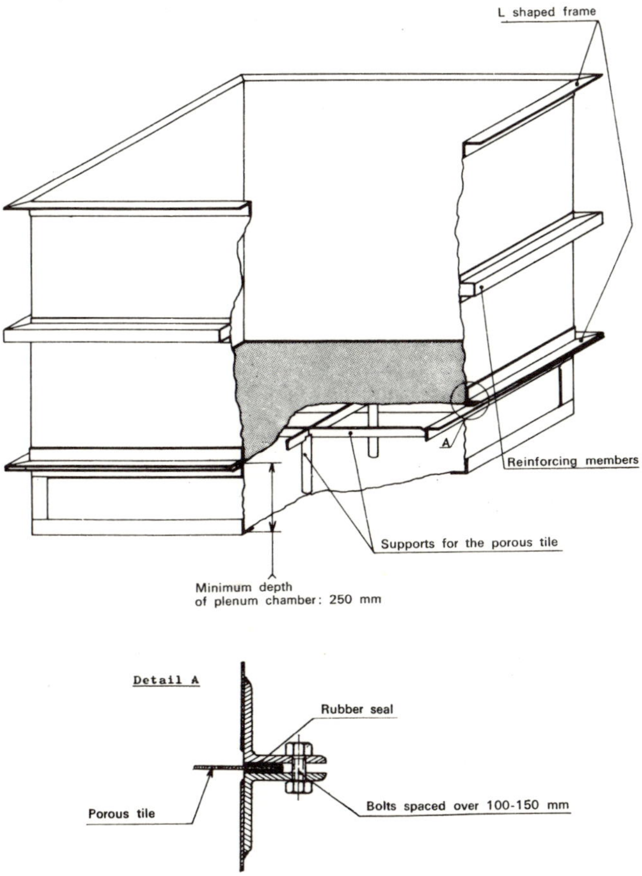

Fig. 90. The fluidized bed

must provide clean air at approximately sixty cubic metres per hour per square metre of porous wall and at a pressure which will be almost directly proportional to the height of the powder within the bed; thus a powder height of 500 mm will require a pressure of approximately 600 mm water gauge.

A butterfly valve is usually fitted to the pipe joining the blower to the plenum chamber so that the output of air may be regulated.

A filter must be placed in this line, and the cross sectional area of the coupling between the pipe and the plenum chamber must be equated to the area of the blower output valve, to minimize the loss of pressure.

The tank itself is usually constructed of mild steel with a framework strong enough to support the tank and stable load. A rubber gasket is used to seal the porous base to the frame and this must fit perfectly to avoid air leaks.

Irrespective of the size of the fluidizing tank the plenum chamber must have a minimum height of 250 mm.

Casters are generally fitted to the tank to permit mobility and a lid should be provided to prevent airborne contamination when the tank is not in use.

The metallic article to be coated is thoroughly cleaned by one or more of the methods described in the section on pretreatment and a primer may be applied to improve the adhesion of the dipped thermoplastic powder coating. When the surface preparation is completed the article is heated in an oven to a temperature which may be calculated from the melting point of the powder, the shape of the article and the thickness of film required. The heated article is then dipped into the fluidized bed and the powder melts in contact with the hot surface. Thickness of the film will be largely dictated by the preheat temperature of the article and the duration of immersion.

The major factors governing the success of the dipcoating operation are: the shape and thickness of the metallic articles, the type of metal, the preheat temperature and the duration of immersion in the fluidized bed.

As a general principle, the coating thickness will depend largely upon the preheat temperature, but the other factors mentioned can have an influence on the uniformity of the film. Where the article is constructed of widely different metal thicknesses, the lower thermal capacity of the thin sections will not allow them to retain heat as effectively as the massive parts. If the temperature is elevated to ensure uniform coating of the thin sections then the heat retained by the massive sections could result in excessive coating thickness or even thermal degradation of the coatings. Dissimilar metals will also have different thermal capacities and an adjustment factor must be applied to cater for both the shape of the article and the metal composition.

A.T.O. Chimie, the well-known manufacturers of Rilsan nylon powders, have suggested the following correction factors, to be calculated and used in connection with the nomograph shown in fig. 91, to obtain the correct preheat temperature and time for coating articles with Rilsan.

Table 17. The Shape Factor

Object	Factor
Tubes for exterior coating only, measure the wall thickness	Multiply by 1·5
Hollow containers, for example fire extinguishers, measure the wall thickness	Multiply by 1·2
Square profiles, wire or iron rods, measure the width of the flat side, or the diagonal diameter	Multiply by 0·4
Plain or embossed sheets, T or U shaped profiles to be coated both sides, tubes coated interior and exterior	Multiply by 1·0

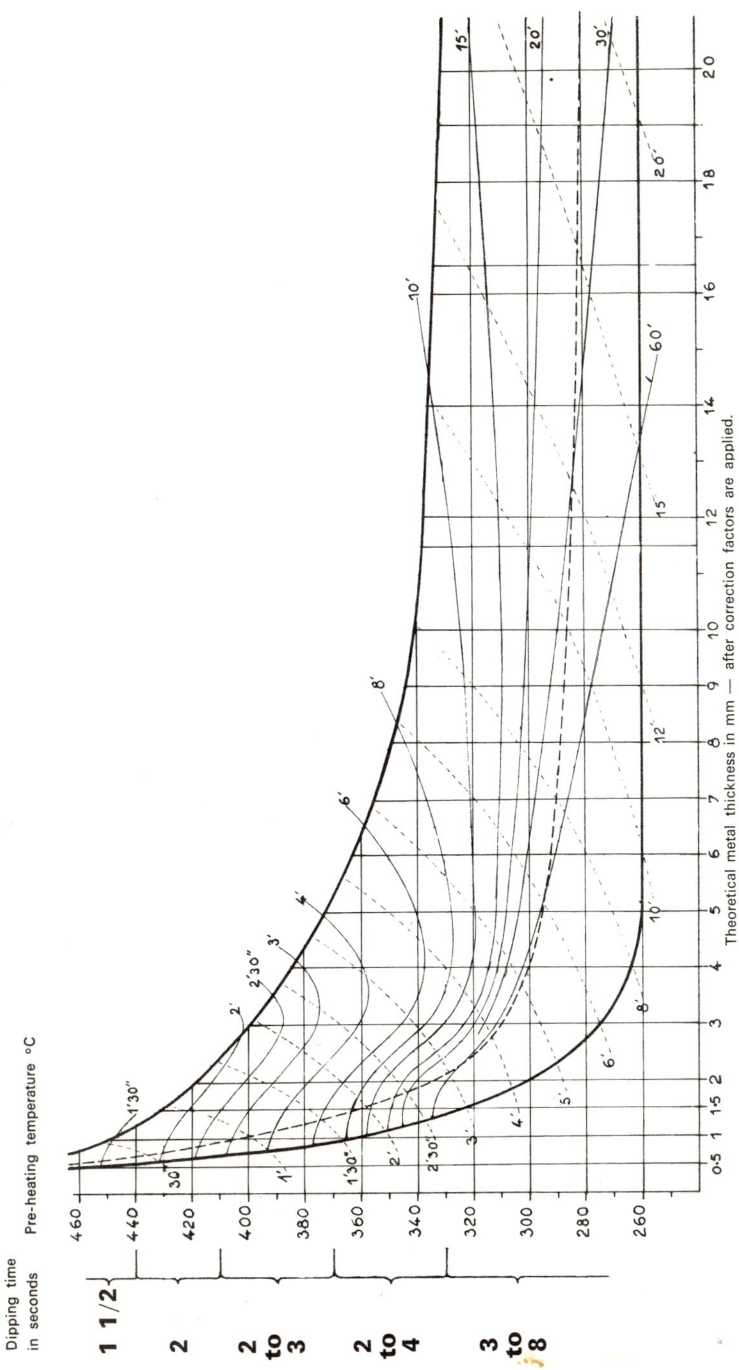

Nomograph to determine the pre-heating conditions of items to be coated by dipping in a fluidised bed of RILSAN Fine Powder.

Fig. 91. Nomograph (Courtesy A.T.O. Chimie)

Table 18. The Metal Factor

Metal	Factor
Aluminium and its alloys	Multiply by 0·7
Copper and its alloys	Multiply by 0·9
Steel	Multiply by 1·0

Both the above factors must be applied to calculate the theoretical thickness.

EXAMPLE

An item to be coated comprises a steel tube and an aluminium sheet. The coating must cover both sides of the sheet but only the exterior of the tube.

Wall thickness of the tube 2·5 mm
Correction factor for shape $\times 1{\cdot}5$
Correction factor for metal $\times 1{\cdot}0$
The theoretical thickness of the tube is $2{\cdot}5 \times 1{\cdot}5 \times 1{\cdot}0$
$= 3{\cdot}75$ mm.

Thickness of the aluminium sheet is 1·5 mm
Correction factor for shape $\times 1{\cdot}0$
Correction factor for metal $\times 0{\cdot}7$
The theoretical thickness of the sheet is $1{\cdot}5 \times 1{\cdot}0 \times 0{\cdot}7$
$= 1{\cdot}05$ mm.

The theoretical thickness can now be used, in conjunction with the nomograph (see fig. 91) to establish the correct heating times and temperatures.

The nomograph shows the following:

1. The theoretical thickness as abscissa.
2. Oven temperatures as ordinates.
3. Times of immersion in the bed on the left of the table.
4. The curves show:

   (a) Heavy broken line. This is the optimum condition line used to establish the preheat temperature from the theoretical thickness. The minimum theoretical thickness is used to select the preheat temperature since this will require the highest temperature;

   (b) Fine continuous lines show the maximum heating times in minutes;

   (c) Fine broken lines show the minimum heating times in minutes;

   (d) Heavy continuous lines show the limits of the possible processing conditions.

Referring to the theoretical thickness calculated in the example the preheating temperatures may now be defined.

At 1·05 mm thickness:
Optimum temperature 400°C
Maximum temperature 450°C
Minimum temperature 360°C.

At 3·75 mm thickness:
Optimum temperature 305°C
Maximum temperature 385°C
Minimum temperature 265°C.

Since the preheat temperature is determined by the thinnest section of the article to be coated it may now be decided that the optimum temperature should be 400°C and the minimum temperature 360°C. The thicker section, however, must not exceed 385°C, so this figure is chosen as the preheat temperature.

Now that the preheat temperature is calculated at 385°C the nomograph can be used to read the maximum and minimum preheating times.

At 1·05 mm thickness:
Maximum preheat time 5 minutes
Minimum preheat time 1 minute 15 seconds.

At 3·75 mm thickness:
Maximum preheat time 2 minutes 45 seconds
Minimum preheat time 2 minutes 30 seconds.

The thickest section must determine the preheat time since it cannot be allowed to over-heat and the processing conditions may now be fully defined.

Heating time	2 minutes 40 seconds
Heating temperature	385°C
Immersion time	2 to 3 seconds.

This method of calculating the processing conditions can only be used effectively if the oven is calibrated correctly. The most practical method of oven calibration is the use of a test piece comprising rods of different thicknesses. When the oven has attained a set temperature, the test piece is placed in the oven for a fixed period and then immersed in a bed of powder for an exactly measured time. Upon cooling, the test piece is examined and the oven may then be adjusted by the further application of this trial and error method. The preheat time and the time of immersion must be accurately measured so that the only variable with each successive test is the oven setting.

The process will give good reproducible results in production if the article to be coated is immersed once only in the fluidized bed for the time necessary to provide the coating thickness. A timer is used to control durations of immersion which are not normally longer than ten seconds. For trial dipping, or to prevent the retention of air which could be entrapped in thick coatings, the necessary film thickness may be obtained by several immersions of one to two seconds duration. The method chosen depends largely upon the shape and complexity of the article.

There is no general rule that can be applied to the dipping of complex shaped articles, although correct jigging and the movement of the article within the bed have an important influence upon the success of the coating operation. A few simple test trials should enable the operator to decide upon the best method of jigging the article so that one of its leading edges is first introduced into the bed; this avoids excessive collection of powder on flat surfaces lying parallel to the surface. Cylindrical articles are best coated by rotating the cylinder during immersion along the axis parallel to the surface and continuing this rotation during cooling. The build up of powder on the sides of certain articles, opposite to the air flow in the bed, may be removed by gently blowing these areas with compressed air as the article is removed from the fluidized bed.

The method of cooling the article will have a marked influence upon the subsequent adhesion and performance of the coating. If cooled too rapidly, internal stresses can arise within the coating, so that it is advisable to allow the article to cool slowly away from draughts. Rapid water quenching of nylon coated articles has been employed to produce a glossy surface but this process can only be used effectively when the article has previously been coated with a suitable primer to enhance the adhesive qualities of the nylon coating.

Sections of certain articles need to be masked-off to prevent the dip coating blocking threads or precisely dimensioned holes. Adhesive tape may be used and the traces of adhesive remaining after the removal of the masking tape can be wiped away with solvent.

Pincers with Teflon coated jaws may also be used to grip the articles in those areas where masking is required.

The methods used for holding articles in the fluidized bed must be carefully considered so that heat losses are minimized, and the points of contact with the holder are not so large in area that the coated article needs extensive touching-up. Grip-tight devices will also prevent accidental dropping of the article into the tank.

Types of Fluidized Bed Powder Coatings

Thermosetting types of powder coatings may be easily comminuted to a particle size range below 100μ which is best suited to application by electrostatic spray methods. Thermoplastic materials are almost impossible to comminute to the small particle size required for spray application without the use of expensive cryogenic grinding techniques and they are therefore used extensively in the fluidized bed method of application.

Particle shape is an important requirement for fluidized bed coating for an incorrect shape will result in poor fluidization, causing uneven films and difficulties in the coating of light gauge wire articles.

The thermal properties of the thermoplastic material must allow an acceptable differential between the melting point and the degradation temperature. It would otherwise be impossible to coat articles composed of different gauge metals, without degrading the coating on the heavier sections while attempting adequately to coat the lighter gauge components. A coating resin should not possess a high degree of crystallinity, for such coatings tend to shrink on cooling and give inadequate cover on sharp edges. The continuous film must also be able to give the properties required by the customer and these may include, exterior durability, toughness, surface hardness, adhesion, good colour and corrosion or chemical resistance.

Nylon 6 and Nylon 66 are unsuitable for fluidized bed coating because they have sharp melting points followed by a rapid reduction of melt viscosity, so that an article coated with these nylons at a temperature only several degrees above the melting point would show pronounced sagging. Thermoplastic acrylics of suitable melting range are generally found to be extremely brittle.

The resins used in large commercial quantities are Low Density Polyethylene, Polyvinyl Chloride, Nylon 11 and Nylon 12. Smaller quantities of High Density Polyethylene, Cellulose Acetobutyrate and Penton are used in special applications.

POLYETHYLENES

The major problem with the use of low density polyethylene is its poor resistance to stress cracking and this defect has prevented its wider application in the fields of chemical resistant and exterior-quality coatings. Nevertheless, the ease of application of low density polyethylene has made it entirely satisfactory as a low cost coating for a range of wirework products where a decorative finish is the main requirement.

Low density polyethylene has undergone considerable development to overcome its known defects, and limited success has been obtained with a range of copolymers with acrylic and vinyl esters. These include the copolymer of ethylene with vinyl acetate (EVA) which is resistant to stress cracking but is difficult to comminute to the correct particle shape.

Copolymers with acrylic acid exhibit excellent adhesion properties, and blends of low density polyethylene with polyisobutylenes and other rubber-like polymers have been developed.

If the decorative appearance of low density polyethylene is not important for the end use application, it is possible to use material with a low melt flow index which has a reduced tendency to stress cracking. A typical application is the coating of underground water pipes.

Polypropylene has been assessed for suitability in powder coatings but the highly crystalline nature of this material causes problems of shrinkage.

The low cost of polyethylene is a major factor in its widespread use but it also gives coatings with high dielectric strength comparable with the more expensive epoxies.

POLYVINYL CHLORIDE

This versatile material is widely used in the United States to give a range of coatings from the soft and flexible to the hard, tough and abrasion resistant finish. The wide range of physical properties is produced by the use of plasticizers and stabilizers. A major fault of PVC coatings is their pronounced tendency to thermal degradation, for the degradation temperature is close to the melt flow temperature. It is not possible to heat PVC coatings to much higher than 180°C even when a heat stabilizer is incorporated. Furthermore the products of degradation include hydrogen chloride which not only promotes corrosion in exposed steel structures in the vicinity but is also a health hazard.

PVC in the unplasticized state is a hard and brittle material so that for most coating applications a plasticizer is an essential component. There is a tendency for external plasticizers to be migratory and to sweat out of the coating and evaporate with eventual embrittlement.

This resin is unsuitable for exterior durability since it has poor chalk resistance comparable to the epoxy resins.

Graft polymers of polyvinyl chloride and acrylic monomers have produced a range of internally plasticized polymers which do not suffer from migration of plasticizer.

Vinyl coatings exhibit good flexibility, high acid resistance and excellent water resistance.

NYLONS

Nylon 11 is a synthetic superpolyamide manufactured by A.T.O. Chimie and sold under the trade name Rilsan. It is produced from castor oil by conversion of the ricinoleate esters to methyl undecylenate. This substance is transformed by hydrolysis and amination to 11-amino undecanoic acid which is poly-condensed to give Nylon 11 with the basic chemical formula.

$$H[HN-(CH_2)_{10}-CO]_n OH$$

The Nylon 11 coatings have excellent solvent resistance, high abrasion and wear resistance, good gloss retention and provide a smooth attractive finish.

It possesses a melting point of 185°C which is slightly higher than the melting point of Nylon 12 at 176°C.

Nylon 12 is manufactured from butadiene and has properties comparable to those of Nylon 11. The market for nylon powder coatings is dominated by the Nylon 11 products.

CHLORINATED POLYETHERS

These are marketed under the trade name Penton and are used for applications where the coating must possess excellent chemical resistance. They have a toughness and film hardness similar to vitreous enamels but with better resistance to impact. Coatings may be machined and they therefore find some outlets for the coating of moving machine parts. A restriction upon the widespread use of Pentons is their high cost.

CELLULOSE ACETO-BUTYRATE

This thermoplastic material gives films with good gloss, colour and colour retention. It is a relatively low cost coating which exhibits excellent free-flowing properties, smooth surface and better heat stability than the vinyl coatings.

CAB products with high butyryl contents may be used as modifiers for other thermoplastic resins since they have good compatibility with many low molecular weight resins.

EPOXIES

The properties of these resins have been described in detail in an earlier chapter. They may be applied by fluidized bed techniques and, with the incorporation of a suitable hardener, they produce thermoset coatings of high build, good electrical insulative properties, good resistance to high operating temperatures, excellent adhesion and low shrinkage factors. The epoxies are widely used for the coating of electrical armatures, capacitors, resistors, and pipework requiring a high degree of chemical and corrosion resistance. In comparison with the common thermoplastic materials they are high priced and this has restricted their use in the fluidized bed operation.

FLUORINATED HYDROCARBON POLYMERS

This class of compounds is characterized by its low-speed coefficient of friction (in the range 0·04–0·06) which is lower than that of graphite and molybdenum disulphide (0·09–0·12). Polytetrafluoroethylene has a static coefficient of friction lower than its dynamic coefficient of friction and this property is responsible for its non-stick qualities. The chemical resistance of these polymers is excellent over a wide temperature range and they also possess good electrical stability.

Fluorinated ethylene propylene is the other major polymer in this class and in possessing a high viscosity at its melting point of 290°C it exhibits typical thermoplastic behaviour. Polytetrafluoroethylene, on the other hand, does not show the melt characteristics of a thermoplastic since it changes from a solid to a rubbery gel at the melting point of 327°C.

The high cost of these polymers has restricted the wider use of both PTFE and FEP although the latter is finding increasing outlets in the coating, by fluidized bed, of a considerable range of manufactured products. Chemical resistance and release coatings are applied to valves, armature cores and insulators at a particle size between 170 and 250 microns.

Surface preparation is a most important stage to ensure void free coatings and the sequence may include solvent degreasing, sand blasting and pre-baking at 600–700°C to remove all traces of volatile contaminants. Maximum adhesion is given when the surface is coated with a special primer before applying the polymer.

One technique which has been employed to lower the coating costs is the two stage fluidized bed coating of a standard thermoplastic polymer followed by the application of a specially designed PTFE powder.

Resins which have been used in this bilayer system include chlorinated polyethers, and PVC.

SPECIAL TECHNIQUES IN APPLICATION BY FLUIDIZED BED

The Continuous Coating Process (see fig. 92)

The fluidized bed may be adapted for the continuous coating of wire since the equipment is considerably cheaper than extrusion coating devices.

Wire or strip is preheated by an induction heating method to reduce the length of the preheating zone (the wire is moving at speeds in excess of seventy metres per minute). The entry and exits points at which the wire passes through the fluidized bed, have guide plate and powder traps fitted to contain the powder within the bed. A water tank is the most effective method for quenching and setting the thermoplastic coating applied to the wire. Tension upon the wire is maintained at the various speeds of operation by a traction device.

Adhesion of the thermoplastic coating to the wire can be improved by the use of a suitable primer, and this may be applied prior to the preheating stage by dipping or, better still, passing the wire through a synthetic sponge material soaked in primer.

The Electrostatic Fluidized Bed

In the next chapter the electrostatic spray application of powder coatings is discussed in detail, but the use of electrostatic techniques in the fluidized bed is properly included in this section.

This method was developed by S.A.M.E.S. in France to improve the efficiency of the fluidized bed and to enable articles of simple shape to be coated, without the need to immerse the article in the bed of powder. The main advantages of this new method were the improvements in cleanliness of the process, and the considerably reduced volume of powder kept in stock within the fluidized bed.

The bottom of the tank consists of a ceramic or porous plastic plate through which air passes to fluidize the powder. Electrodes connected to a high voltage generator are distributed over the porous plate. When an earthed article passes over the powder an intense electrical field is created

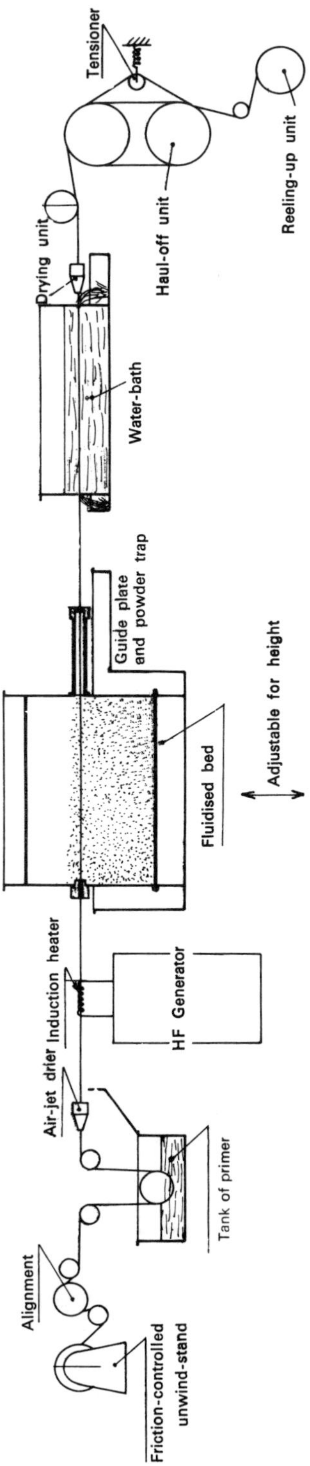

Fig. 92. Continuous wire coating line

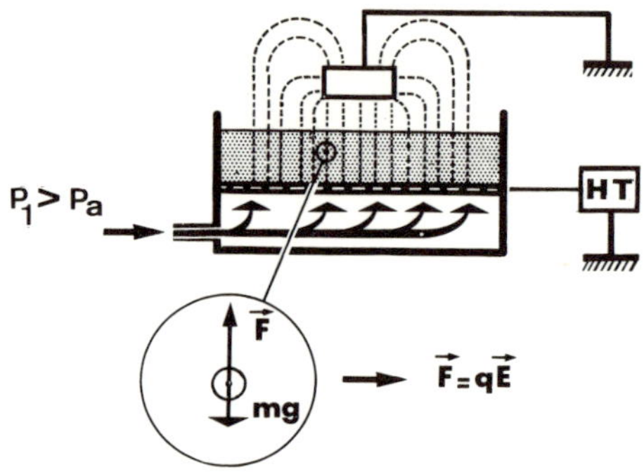

Fig. 93. Electrostatic fluidized bed

between the article and the electrodes. Powder particles receive an electrostatic charge and they are conveyed to the earthed article by the transporting influence of the electrical field.

It is almost impossible to coat intricate articles by this method but it has proved to be successful with small objects, continuous wire and webbed material. Film thicknesses of up to 500 microns may be obtained and it is not necessary to preheat the article to cause the powder to adhere, although the objects must be heated after coating to flux the powder layer and produce a continuous and fully adherent film.

The electrostatic generator supplies d.c. voltage which is variable between 30 and 90 kV. A control panel incorporates the generator, together with variable adjustment of the pressure required to fluidize the powder.

Safety requirements for the electrostatic fluidized bed are more stringent

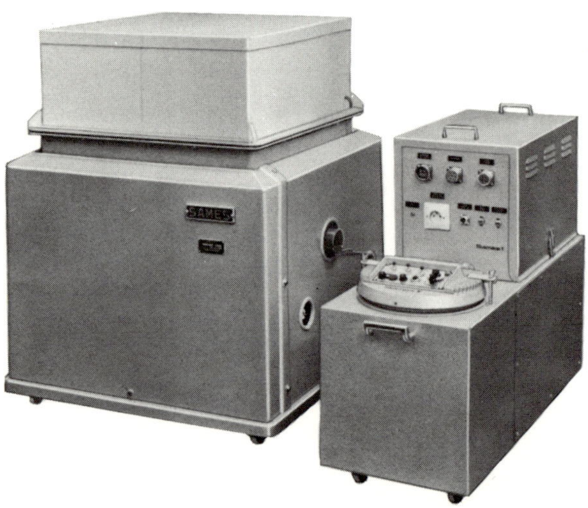

Fig. 94. The Stafluid bath (Courtesy Binks Bullows Ltd.)

than those applied to the conventional fluidized bed and for this reason, the manufacturers will only supply this equipment for fully automated production lines where the vital safety precautions can be implemented.

REFERENCES

1. Erwin Gemmer *et al.* (Knapsack-Griesheim).
2. British Patent 862494 (March, 1961).

10

Powder Coatings Applied by Electrostatic Spray

SUMMARY

In 1962, the French spray equipment manufacturers S.A.M.E.S., developed the electrostatic powder gun as a logical extension to their activities in the field of electrostatic wet paint spray application. Since that time, the powder process has made a world-wide impact upon the traditional industrial paint markets with a meteoric growth rate that has been relatively unaffected by a general down-turn in overall economic expansion.

The electrostatic powder process has been investigated by a number of academic institutions, notably the University of Southampton under the direction of Professor Bright, and the practical application of the basic principles of electrostatic charge acceptance has resulted in the development of a wide range of spray equipment, with manufacturers' each claiming additional advantages derived from the use of their equipment. There is no doubt that during this period of innovation many of the claims made by equipment manufacturers are justified as logical improvements to the basic concept.

These claims, and the manner in which they relate to the theories have understandably caused some confusion, mainly due to the fact that powders, as developed at the present time, do not exhibit idealized electrostatic behaviour.

In this chapter the basic ideas are reviewed with emphasis placed upon the properties of powder coatings and the way in which these properties affect the deposition efficiency.

The properties include particle size, particle shape, surface resistivity and the chemical composition of the powder. Efficiency of deposition will also depend upon the charge acquired by the particles, the velocity of the transporting air, the viscous drag of the air and the distance the particle has to travel.

Deposition upon the coated article will depend upon the size and shape of the article, the air extraction rate within the booth, the film thickness required and the resistivity of the surface.

There are several important aspects of the electrostatic deposition of powders which are highlighted in this chapter. These include:

(a) Spherical particles accepting charge more readily than irregular shaped particles.

(b) Larger particles being more efficiently charged than smaller particles and

depositing more readily on the article with the fines tending to accumulate in the overspray.
(c) The development of a self-limiting film. It is only possible with powders of high resistivity and the film thickness, at which this self-limiting mechanism operates, is in excess of the film thickness that is normally commercially acceptable.
(d) Powders of medium resistivity exhibiting the phenomenon of back-ionization if the gun is held too close to the article. This can result in pitting or starring of the surface.

CHARGE ACCEPTANCE OF POWDERS

In the influence of an electrostatic field all particles become electrically charged due to the loss or gain of an electron or molecular ion. They become polarized by the formation of a dipole resulting from the orientation of their molecules or the displacement of their free electrons.

An electrically charged field may be obtained by several different mechanisms and three methods have been used to charged powder coatings.

Corona Discharge

This is the most widely used mechanism and the electrical field results from the discharge of gaseous ions at a point source of high electrical energy.

When the surface field intensity at the point source exceeds 3 volts/micron, a corona discharge can occur, and in practice this is achieved by maintaining a potential of 100 kV at the needle point mounted in the spray nozzle.

The charge acquired by the particles is derived from Pauthenier's equation.

$$q = 12\pi\varepsilon_0\left(\frac{\varepsilon}{\varepsilon + 2}\right)E_c a^2$$

where q is the charge in coulombs,
ε_0 is the permittivity of free space,
ε is the dielectric constant of the powder which is approximately 2,
E_c is the electrical field at the charging point which has been measured by Corbett to be in the region of 10^6 V/m,
a is the mean particle radius.

For a particle size of 50 microns it can be calculated that the particle charge is of the order of 10^{-13} coulombs.

A simple expression for measuring charge efficiency is the charge/mass ratio

$$q/m = \frac{9\varepsilon_0 E_c}{2a\rho}$$

where ρ is the density of the powder expressed in kg/m^3.

At 50 microns and a density of 10^3 kg/m^3, q/m is approximately $1{\cdot}8 \times 10^{-3}$ coulombs/kg.

The sprayed powder may be collected and the total charge and mass measured to give a practical method of assessment.

Charging by Induction

Particles with low resistivity cannot hold their charge when deposited on a conducting surface, but, since they are still subjected during application to an electrical field, it is possible to reverse their charge by induction. This effect can result in the removal of the induced charged particle from the deposited coating.

An equation similar to that defining acceptance of corona charge can be applied, although the extent of charge acceptance will inevitably be lower.

Tribo-electric Charging

This phenomena may best be described as friction charging. The rubbing together of powder particles of high resistivity in a field of low electrical intensity can produce a charge acceptance figure which approaches that obtained by corona discharge. A high relative humidity can profoundly effect the efficiency of triboelectric charging, but this is probably due to the effect that high humidity has upon the powder resistivity.

The important feature of this mechanism is that no high voltages are required to arrive at a current density comparable to that obtained by corona discharge.

TRANSPORTATION OF THE CHARGED POWDER

Charged particles leave the electrostatic gun and proceed towards the earthed article under the influence of three forces.

(i) The airflow from the gun, which was used to convey the powder from the feed hopper to the gun nozzle now carries the charged particles towards the article. An excessive stream of air will result in a large amount of overspray which will reduce the deposition efficiency. The operator should not increase the air feed to the electrostatic gun in order to obtain better penetration of powder in enclosed sections of the article, but should rely upon the electrostatic attraction to achieve a uniform coating. The composition of the oversprayed powder will largely depend upon the velocity of the conveying air stream. An examination of the equation relating to acceptance of corona charge, shows that the larger particles receive a more efficient charge which selectively deposits the large particles while the small, less efficiently, charged particles accumulate in the oversprayed material. A high velocity air stream, however, would increase the momentum of the large particles and convey them past the article, thereby reducing coating efficiency.

The segregation of particle sizes during the spray operation is one reason why the powder manufacturer recommends that oversprayed powder is mixed with fresh powder prior to re-use.

(ii) An electrical force will drive the particle towards the earthed object and the velocity attained depends upon the electrostatic field and the viscous drag of the air.

The velocity is defined by.

$$V_E = \frac{\varepsilon_0 E_c E a}{\eta}$$

where E is the mean electrical field surrounding the particle and η is the coefficient of air viscosity.

For correct application the velocity V_E must be much greater than the extraction velocity of air in the coating booth. The mean electrical field between the gun and the object is of the order 10^5 V/m. An average velocity V_E is around 1·5 metres/sec. if the particle is correctly charged.

The equation also shows that in an electrostatic field fine particles move more slowly than large particles so that uniformity of particle size is a desirable feature in a continuous coating process.

Corbett[1] has reported some evidence that the electric field in the vicinity of the powder layer on the article is greater than the electrical field between the gun and the object. This value could be important in assessing the impact velocity of the powder and the effect of electrostatic removal forces on conducting powder.

(iii) A gravitational force will also effect the transportation of the powder particles and the effect will be more pronounced at higher particle sizes. Calculation shows that the gravitational effect will only become a significant factor with particles with diameters greater than 100 microns.

THE DEPOSITED LAYER OF POWDER

In the period after deposition of the powder and before the stoving operation, the powder layer adheres to the surface by electrostatic forces, Van de Waals intermolecular forces and packing brought about by the application of these cohesive forces. Powders will tend to be dislodged by the mechanical conveyance of the object, gravitational and air extraction forces, and, in the case of conducting powder, the electrostatic removal effect.

It has been reported by Bright[2] that the decay of charge in the powder layer can be calculated from

$$q = q_0 e^{\frac{-t}{\rho \varepsilon_0 \varepsilon_r}}$$

where q_0 is the initial charge, ε_r is the dielectric constant, ρ is the resistivity.

Most organic powders have high resistivities, greater than 10^{13} ohm/m and the half decay period is in the region of three minutes.

It is extremely difficult to remove charge from particles of high resistivity and this makes the collection of oversprayed powder in a cyclone a potential explosion risk due to tribo electrification mechanisms.

In the intermediate resistivity range 10^9–10^{12} ohm/m an interesting phenomena has been reported by Bright. If the product of the resistivity and the ion current density in the electrical field within the powder layer is in excess of 10^6 then back ionisation can occur resulting in starring of the surface.

This effect is observed in metallic pigmented powder coatings and some high quality carbon blacks, and the defect can only be overcome by increasing the resistivity of the carrier.

Powder coatings with low resistivity, below 10^8 do not exist in the current organic range and the comments relating to their deposition efficiency are not relevent to the subject matter of this chapter.

Particles with resistivities below 10^9 require high ion current densities for deposition and must be of uniformly small size. The thickness of film is not limited by the repulsion of incoming particles due to the electric charge in the powder layer.

In the resistivity range 10^9–10^{12} ohm/m the film is not self-limiting and moderate ion current densities are required. If the gun is held too close to the object then back ionisation will occur with pitting of the stoved film.

High resistivity powders will be self-limiting and optimum build will be obtained within a few seconds. A low ion current is required and particle size and spraying conditions are not critical.

The limiting thickness is, however, rather greater than that required by most end users and application procedures are seldom designed to take advantage of this feature. It may be that the self-limiting property can only be utilized with uniformly smaller particles of higher resistivity.

The self-limiting effect is not observed at sharp edges where excessive build can occur, or in inaccessible sharp corners which are starved due to the so-called "Faraday Cage" effect.

Perhaps it is possible that the self-limiting effect is evidenced more quickly in sharp corners, for the author has observed that an extremely thin film of powder is rapidly deposited and the slow dissipation of charge could prevent further penetration. If this observation is correct then the mechanism of tribo electrification should overcome the "Faraday Cage" effect.

Golovoy has presented a critical evaluation of the deposition efficiency of various acrylic powder coatings.[3] This investigation considered the following variables: substrate size, velocity of air extraction in the booth, film thickness, charging voltage, spraying distance, particle size, shape and surface resistivity.

The following results of his evaluation show the effect of each of the above variables upon the deposition efficiency.

Substrate size will influence deposition efficiency, in proportion to the width, for most elongated articles. The ratio of the width W to the outlet diameter of the spray gun D is expressed by W/D and an increase from 0·5 to 3·0 in the W/D ratio was shown to improve the deposition efficiency logarithmically from 20% to 65%. Further increase in this ratio resulted in a limiting value being attained.

The influence of the booth air velocity is predictable and Golovoy showed that little or no effect upon deposition efficiency was evident at air flow speeds up to 3 feet/sec. parallel to the direction of spray.

Higher air speeds would have an adverse effect upon deposition efficiency but these were not normally encountered with commercially available booths.

Air flow speeds in the direction normal to the powder flow were shown to have a marked effect upon deposition efficiency although this adverse factor could be reduced by decreasing the spraying distance.

Film thickness up to 1–2 thou (25–50μ) at a charging voltage of 60 kV does not effect deposition efficiency.

As the film thickness increases above this level the deposition efficiency decreases linearly with the thickness of deposited film. The effect of increasing the charging voltage is to increase the rate by which the deposition efficiency is reduced.

Generally, an increase in charging voltage results in improved deposition efficiency, although this effect is less pronounced with increasing film thickness. An increase in charging voltage increases the average charge on the particles, and the higher electric field provides greater electrostatic transportation forces. These two effects, are off-set to some extent by the greater residual electrostatic field in the deposited layer. It was reported that a step-wise increase in charge voltage as the film thickness increases does significantly improve the deposition efficiency.

Spraying distance is a factor which can affect deposition efficiency especially at low charge voltages, due to its influence upon the electrostatic transporting field. The decrease in efficiency with increased spraying distance is not so pronounced at higher voltages and this may be related to the dissipation of charge from the deposited particles into the surrounding electrical field. Applying the film in a multicoat operation by increasing the distance and the voltage with successive coats gave a deposition efficiency of 75% at 3·5 mils.

An interesting aspect of Golovoy's investigation is the effect of particle size and shape. The deposition efficiency increases rapidly with increase in particle "diameter" up to 35 microns. At particle sizes above this level the efficiency decreased due to difficulties in fluidizing and conveying larger particles. Fine particles were greatly affected by the air flow and the deposited film was found to be composed of particles having an average size greater than that of the powder before spraying.

Spherical particles gave deposition efficiencies of 70% compared with 40% for irregular shaped particles.

In the earlier discussion of surface resistivity it was stated that particles with a high resistivity retained their charge for a longer period than particles of low resistivity. A reduction in the surface resistivity will diminish the adverse effect of the field due to the deposited particle, and improve deposition efficiency.

An acrylic powder whose surface resistivity was reduced from $1{\cdot}3 \times 10^{16}$

to 2×10^{14} ohm/metre showed an increase in deposition efficiency from 40% to 65% at 2 mils film thickness.

The adhesion of the powder deposited is dependent on the residual charge so that poor particle adhesion to the substrate will occur at a surface resistivity below 10^8 ohm/metre.

Golovoy's results briefly summarized show that the deposition efficiency can be increased by:

(a) An increase in charge voltage and a reduction in the spraying distance.
(b) A decrease in the field due to deposited particles
(c) Maintaining the air flow parallel to the direction of spray.
(d) Eliminating fines.
(e) Applying spherical particles.
(f) Adjusting spray diameter to less than one quarter of the width of the article.

THE ELECTROSTATIC GENERATOR

Generation of the high voltage supply is conventionally achieved by a high voltage transformer fed from a single phase 200–250 volt 50 cycle supply. The output from the secondary winding is amplified from around 5 kV to 100 kV by a series of voltage doubling rectifier circuits (essentially Cockroft–Walton multipliers). High voltage output levels are selected by the choice of tapping between the primary and secondary windings of the transformer and the number of stages used in the multiplier.

A major disadvantage of this classical system of high voltage generation is the use of high value capacitors which can store excessive energy. These are not only capable of producing dangerous current levels but they are also bulky and costly.

Modern electrostatic generators overcome this problem by considerably increasing the mains frequency to the voltage multiplier by the use of a high-frequency valve oscillator. The increased frequency, in the range 50,000 c.p.s., permits a drastic reduction in the capacity value of the capacitors in the circuit, with a corresponding reduction in the levels of stored energy. This type of oscillator will cut-out in the event of an overload with instantaneous collapse of the high voltage.

A comprehensive account of the circuitry and design principles of electrostatic generators has been given by Lever.[4]

Supply of power to the output stage of the generator may be governed by three alternative systems of control.

1. In the conventional generator the short circuit load currents will be defined by the internal resistance of the load. When the discharge electrode of the gun approaches the earthed article the voltage falls and the current rises to the level of short circuit allowed by the limiting values of the resistors incorporated in the circuit. The total energy in a

system using resistance control can be as high as 12 watts and it is therefore difficult for this high energy field to penetrate the "Faraday Cage".

A generator with an output of 100 kV and an internal resistance of 500 megohms would produce a short circuit current drain of 200 microamperes.

When the voltage is reduced the current will decrease in a linear manner.

Though this system may be effective at the switch-off point it is suspect during the switch-on operation for the maximum output voltage returns to produce the real possibility of spark discharge.

2. In the second control method a safety cut-out switch is introduced to eliminate the possibility of the current rising to a discharge level. The voltage will still decrease as the gun approaches the earthed article, with corresponding rise in current, but it is possible to select a suitable trip level which is normally between 10 and 35 microamperes.

   A Schmitt trigger circuit of resistors will enable the selected current level to be maintained without the need for the operator manually to reset the circuit.

   There are several disadvantages to the use of a current control switch because during the period when the cut-out is in operation there will be no corona discharge and the operator may be unaware that powder is no longer being applied electrostatically. The switch-on can produce high transient levels of current and energy levels of several joules are possible. There is also the additional possibility of a failure in the transistor switch resulting in a real danger of arcing from the gun.

3. The third method of control employs a current regulator. Voltage fed to the power output is adjusted by a current sensing circuit so that the load current is maintained constant irrespective of variations in the load resistance. This device removes the need to incorporate high resistance elements in the electrostatic spray gun since it controls the current over a wide range of load resistance. It will also protect the power supply against short circuits in the load and limit the short circuit current to a safe level.

   In practical terms it now becomes possible to use low resistances of 20 megohms in place of the 500 megohms or more necessary to safeguard a non-regulated supply. It can be shown that this feature gives much improved voltage regulation so that a voltage of 60 kV at a resistance of 20 megohms will produce an electrical field of similar strength to a voltage of 100 kV from a high resistance circuit.

Levels of energy in the electrostatic field may be calculated from the equation

$$J = \tfrac{1}{2}CV^2$$

where J is the energy in joules, C is the electrical capacitance in farads and V is the voltage.

When a constant current power supply is used the reduction in energy level results from the controlled reduction in voltage. A voltage of 50 kV can reduce to 5 kV as the gun approaches earth, and assuming an electrical capacitance of 100 picofarads the energy reduction will be

At 50 kV

$$J = \tfrac{1}{2}(100 \times 10^{-12}) \times (50 \times 10^{3})^{2} = 125 \text{ millijoules}$$

At 5 kV

$$J = \tfrac{1}{2}(100 \times 10^{-12}) \times (5 \times 10^{3})^{2} = 1{\cdot}25 \text{ millijoules}$$

It is a usual practice to incorporate the safety overload trip into the constant current circuit as an additional precaution against failure of the constant current control. Most systems employing constant current control operate between 75 and 100 microamperes.

While these two safeguards are effective it still allows some energy to be present at short circuit conditions.

A recent development which has now been applied to electrostatic generators for powder coatings, is the introduction of transducer materials into the high voltage circuit. These are input and output devices, which are capable of converting one form of energy into another, and this property is sometimes referred to as the piezo-electric effect.

A piezo-electric material is one in which an electric potential appears across certain surfaces of a crystal while the dimensions of the crystal are being changed by the application of a mechanical force. These materials occur in nature and are naturally polarized. Synthetic materials, however, can be compounded from crystalline powders. For example barium titanate ceramics are made by moulding the ceramic out of a powder of small crystals, baking the ceramic under pressure and placing the material in a strong d.c. electric field.

The compounded crystal is polarized along the axis of the d.c. field direction and exhibits piezo-electric properties.

Synthetic compounds have the advantage that they can be moulded to any shape and size and polarized in the required direction. Compounded layers of these substances distorted to give "twisters" and "benders" are used for a wide range of transducer purposes in the electronic field.

In the high voltage circuit they are able to pass only limited amounts of electric current and therefore provide much lower levels of energy.

The important aspect of the use of transducers is the steady reduction of both voltage and current as the discharge point approaches the earthed article with a complete collapse of both at the short circuit condition. There are no switching mechanisms to control this effect and the energy output is less than 0·5 watt at short circuit.

There are many suppliers of electrostatic spray equipment for powder coatings, and a review is given of the major manufacturers and the essential design features of their equipment.

In the appendix there is a list of all the suppliers and full details of equipment should be obtained direct from the manufacturers listed in this section.

CONTROSION

Electrostatic spray equipment is manufactured in the United Kingdom by Controsion (Electrostatic) Ltd., and they offer facilities for design and layout of manual or completely automated powder coating plants. Testing facilities are also available.

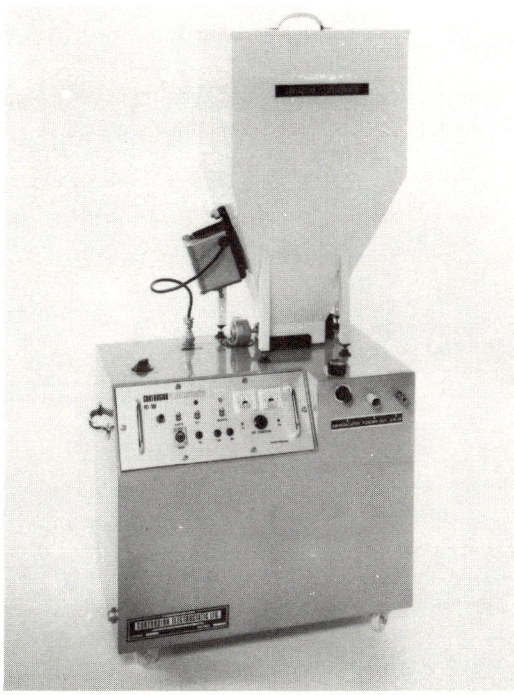

Fig. 95. Complete powder unit incorporating PET-SRF module Model 6000 (Courtesy Controsion (Electrostatics) Ltd.)

The most important aspect of the Controsion equipment is the unique PET-SRF electrostatic generating system.

Controsion have coined the expression "self-regulating field" to describe the reduction of field intensity when the gun head approaches the earthed substrate and the complete collapse of both voltage and current when the discharge electrode is placed into short circuit.

Methods of generating high voltage have already been discussed and mention has been made of the use of piezo-ceramics. These transducing materials allow only limited amounts of current to pass and this fact, coupled with the system in which they have been employed, giving lower energy emission of the apparatus, produces two attractive features in the Controsion equipment.

1. As the head of the gun approaches the earthed article there is a steady reduction of both voltage and current so that a short circuit cannot occur when contact is made, for at this point there is zero power output. This inherently safe working feature and related gun design has complied with the testing procedures of the B.A.S.E.E.F.A. standard SFA 3010 Part 1; 1974 which is discussed fully in the chapter relating to Safety.
2. An exceedingly low energy output of less than 0·5 watt during operation, compared to energy outputs in the region of 12 watts for conventional generators, and the reduction of this low figure close to the substrate, gives an almost total absence of the "Faraday Cage" effect. Articles which have proved to be difficult to coat in certain obscured areas, are adequately coated with the Controsion equipment.

The solid-state transducer material is less prone to deterioration due to heat, and is therefore more reliable.

A PET-SRF generating module measures only $15 \times 5\frac{3}{4} \times 10$ in.

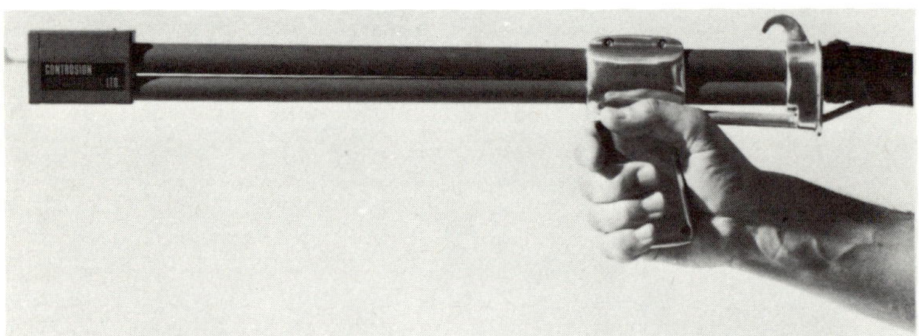

Fig. 96. The FHCC electrostatic hand gun (Courtesy Controsion (Electrostatics) Ltd.)

The FHCC light-weight gun has good transfer efficiency of corona charge and this ensures a high production rate and uniform deposition. Spray pattern may be varied by interchangeable non-stick deflector discs.

The powder feed hopper is interchangeable and it is possible to effect a colour change with a production unit in three minutes by replacing the hopper and blowing air through the feed lines.

Powder within the hopper is not fluidized and it is conveyed to the spray gun by a worm-screw which ensures a constant and reproducible flow of powder from the spray gun. The rate of emission can be accurately controlled and varied by adjusting a d.c. motor rotating the worm-screw.

This method of conveying powder avoids the tendency to "puffing" of the powder when commencing to spray. It also reduces the separation of particle sizes within the powder and prevents build-up in the feed lines and in the injectors.

Fig. 97. Ten gun generator unit based upon five modules (Courtesy Controsion (Electrostatics) Ltd.)

Model 6000 is the complete electrostatic spraying unit for automatic or manual use and has a built-in powder feed system, powder hopper and spray gun.

The PET-SRF generator is available in console form with separate powder feed units and guns for large multi-gun automatic systems. The nominal range of voltage supplied by the generator is 20 to 45 kV with an ionisation current of 10 microamperes at a distance of six inches.

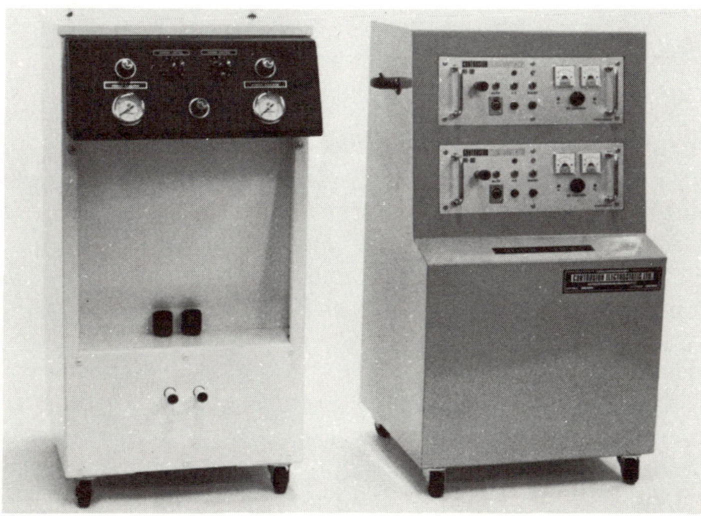

Fig. 98. Model 2 × 6000 unit with 2S × 100 powder feed (Courtesy Controsion (Electrostatics) Ltd.)

A twin gun version of the Model 6000 is also available and is recommended for use in conjunction with the 2S × 100 twin powder feed unit.

GEMA

The Swiss manufacturers of Gema equipment are represented in the United Kingdom by Tycoate Industrial Finishes, who market the Gema range of electrostatic powder spray equipment in conjunction with a complete plant installation service.

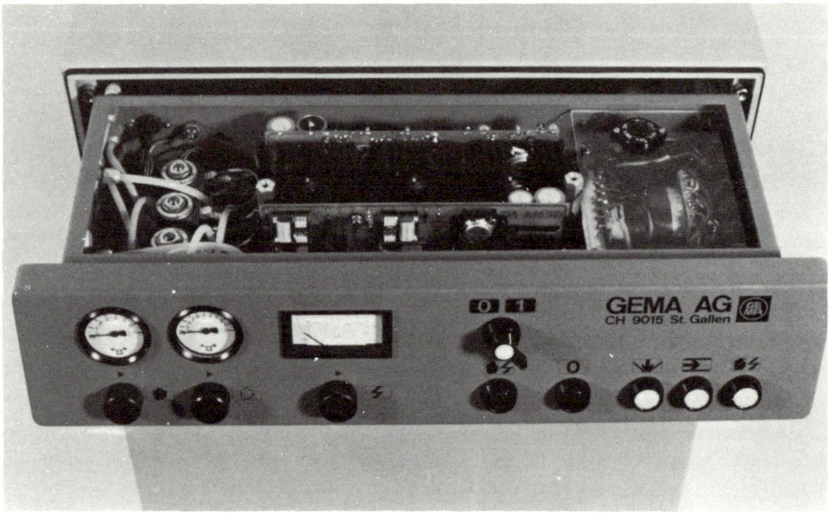

Fig. 99. Modular control unit (Courtesy Gema A.G.)

A major feature of the design in the Gema range is the use of a separate control unit for each gun to provide the necessary electrical supply, and to regulate the pneumatic conveyance of powder from the feed hopper to the gun. This system can be extended to transform a single hand gun set-up into a large automatic unit by merely increasing the number of electropneumatic modules to correspond with the increased number of spray guns.

The hand gun is of simple design and weighs only 790 grams. A high-voltage transformer is set in the gun itself so that high voltage cables are not required between the generator and the gun. This safety feature ensures that no more than 30 V is conveyed between the electropneumatic module and the gun, thereby removing the risk of shock and the considerable danger of spark discharge if the cable is damaged.

Powder entrained in the air stream passes through four electrodes set in the nozzle of the gun so that efficient charge acceptance is assured by the high strength of the electric field.

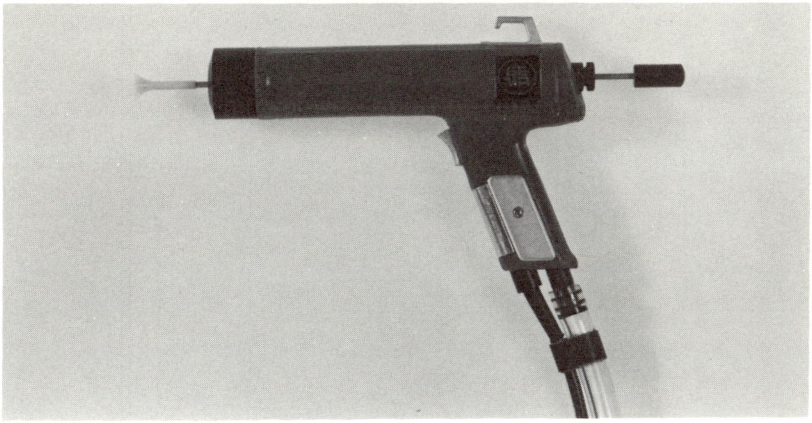

Fig. 100. Hand gun (Courtesy Gema A.G.)

Between the electrodes in the gun, there is an electrostatic field which is substantially independent of the distance between the workpiece and the gun. A further electrostatic field, however, is set up between the gun head and the workpiece which varies in strength as the gun approaches the object.

Although the Gema gun has a relatively low energy high voltage generator, an electronic current limitation is imposed by the inclusion of an automatic cut out, to completely remove the risk of arcing between the gun and the object.

A sliding deflector plate in the Gema manual gun enables the operator to adjust the spray pattern to suit the shape of the article to be sprayed. In the Gema automatic gun, the diameter of the deflector plate is varied to suit the particular application.

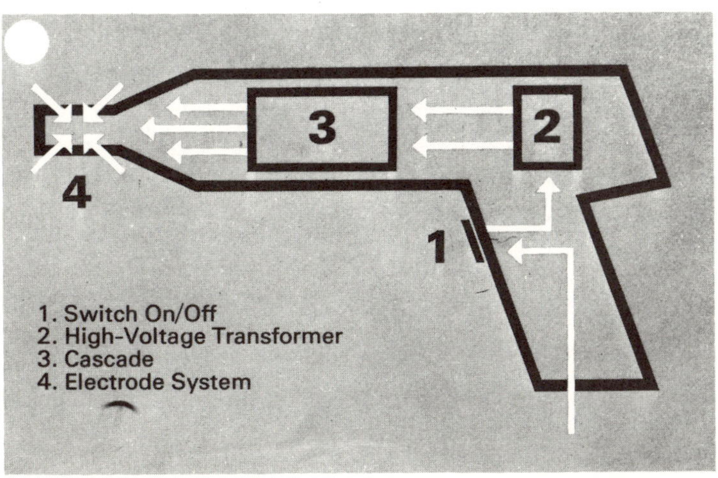

Fig. 101. Schematic layout of hand gun (Courtesy Gema A.G.)

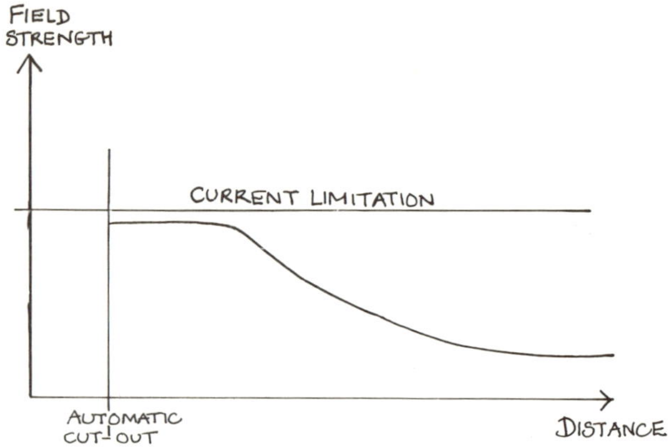

Fig. 102. Graph showing current limitation with increasing field strength (Courtesy Gema A.G.)

Controls on the module allow the operator to alter the rate of the conveying air, the amount of powder supplied to the gun and the applied voltage.

The Gema Range includes:

Type 710 Machine

This is a completely portable machine, primarily intended for laboratory use. The hopper has a capacity of one kilogram.

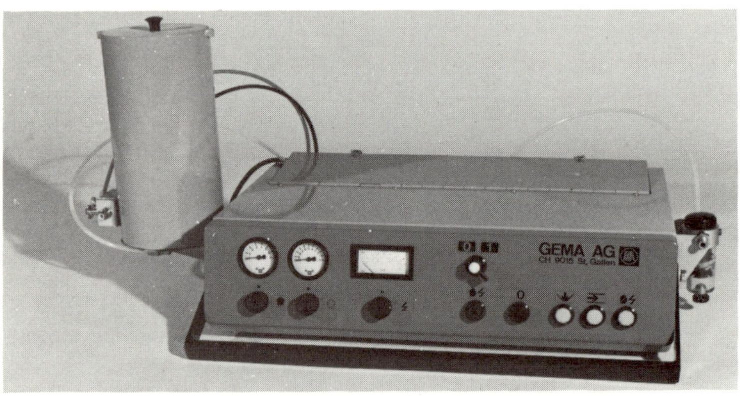

Fig. 103. Type 710 machine (Courtesy Gema A.G.)

Type 721 Machine

The standard model can be either manually or automatically operated. It has a self-contained console incorporating a 25 kilogram hopper. The machine is adaptable to take a second control console and gun at a subsequent stage.

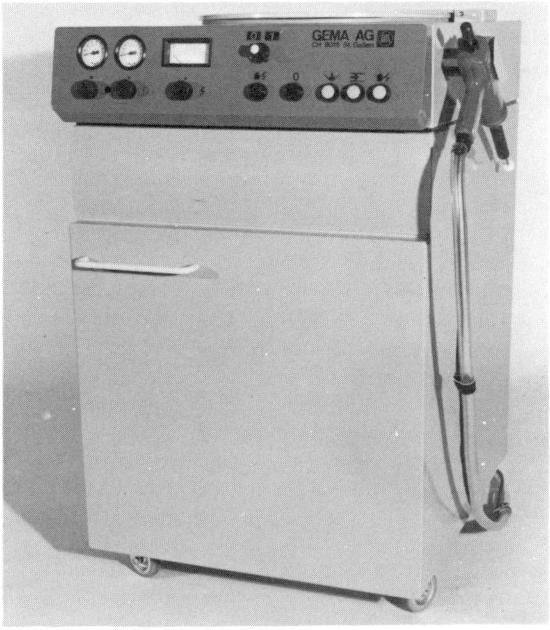

Fig. 104. Type 721 machine (Courtesy Gema A.G.)

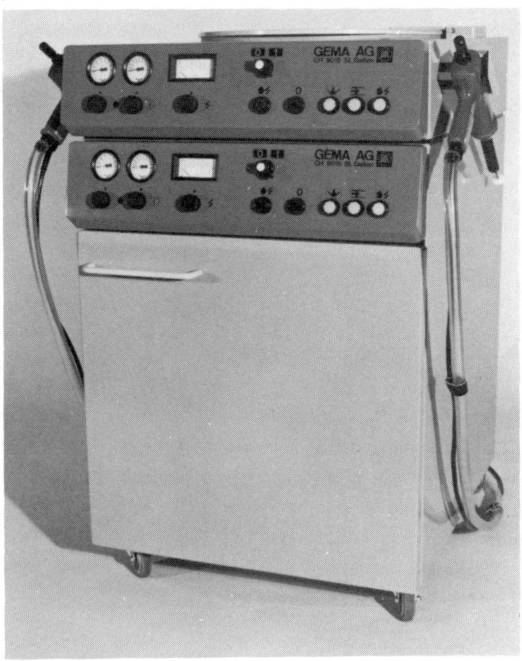

Fig. 105. Type 722 machine (Courtesy Gema A.G.)

Type 722 Machine

This machine feeds two guns, each entirely independent of the other. It is widely used on both a manual and automatic basis.

Type 730 Machine

The standard machine is equipped with a 250 kilogram hopper and is capable of supplying any number of guns up to twelve. Special equipment can be manufactured to accept as many as twenty-four or thirty-six guns. Additional refinements are overriding controls for switching off all guns simultaneously and level controls for the hopper.

Fig. 106. Type 730 machine (Courtesy Gema A.G.)

Multicolour Systems

The problem of quick change between colours is resolved with the Gema equipment by installing a separate hopper and gun for each colour. A lead from a central control console is then simply plugged into the new hopper.

A more sophisticated system is available to switch over controls from a central position to various hoppers electronically. The same system will switch over Enclosures, Recovery Systems, Elevators and all ancillary equipment from one colour to another simultaneously.

Gema Elevator

This device traverses the gun up and down automatically. The length of traverse, the traversing speed and the positions of the reversing points can all be varied by adjustments at the electronic control cabinet whilst the equipment is in operation. The control cabinet can be linked to a stroke limiting device which will alter the stroke automatically to suit the component hanging from the conveyor at any one time.

Gema Elevators are available with the following effective travels as standard, 1·3 m; 1·6 m and 2·1 m. Other travel lengths are available as specials.

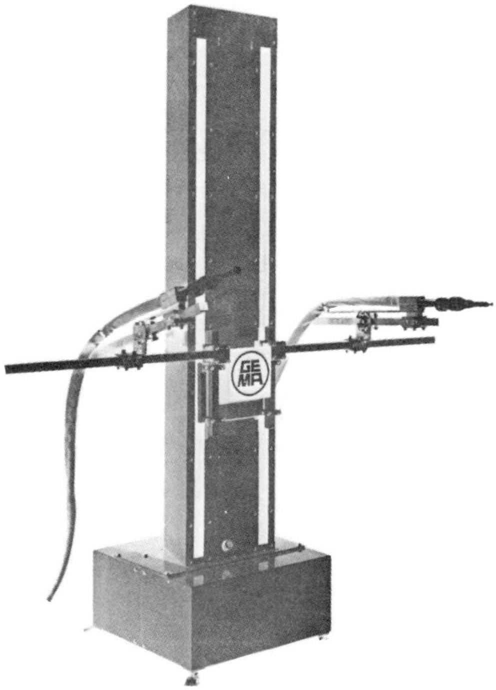

Fig. 107. Gema elevator (Courtesy Gema A.G.)

The standard equipment can be made to traverse either vertically or horizontally as required; a special unit has been designed to follow oblique or curved paths by operating both vertical and horizontal traverse mechanisms simultaneously.

VOLSTATIC

The Volstatic electrostatic powder spraying equipment has been developed and is manufactured by Volstatic Coatings of London, England.

In common with other major electrostatic powder spray equipment manufacturers, Volstatic have developed a comprehensive range of products which are designed to give simplicity of operation, easy maintenance and freedom from working hazards.

The Volstatic generator operates on single phase 200/250 V mains supply at 50 cycles and provides an output voltage of 100 kV. Output current is limited to a maximum of 100 μA.

Two guns may be powered from the same generator unit since the electronic circuits are duplicated.

The internal view of the constant current generator (fig. 108) shows the

Fig. 108. Internal view of constant current generator (Courtesy Volstatic Ltd.)

plug-in modules for trip and current control. Supply frequency to the voltage multiplier is increased by the inclusion of a high frequency valve oscillator.

The electrostatic spray gun weighs only 900 grams and the trigger switch operates at 12 volts. A patented discharge electrode concentrates the electrostatic field in the path of the powder spray and the spray may be adjusted from a direct jet to a circular pattern.

A high-voltage cable between the gun and the generator uses a conducting liquid in place of the normal wire, and this feature eliminates the possibility of electrical discharge if the cable is accidentally damaged.

The laboratory model electrostatic spray gun has a small feed cup, attached to the generator console, which holds about 200 grams of powder. It may be cleaned down in less than two minutes.

The "Autocoater"

The Volstatic "Autocoater" is an integrated automatic powder coating unit constructed from mild steel to comply with safety requirements. It is a completely portable unit which can be delivered ready for use to a site where it could operate with an existing conveyor system.

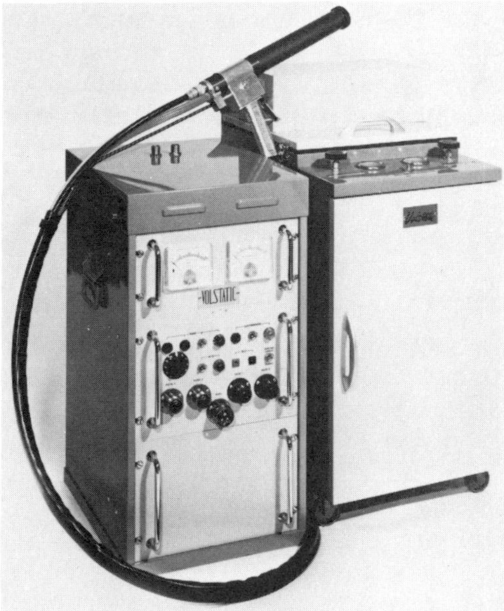

Fig. 109. Powder spray unit with separate hopper and generator unit (Courtesy Volstatic Ltd.)

Fig. 110. Six gun "Autocoater" (Courtesy Volstatic Ltd.)

All power systems, controls, guns and powder feed are built into the fabricated structure of the "Autocoater". The spray chamber is circular in shape with guns fitted to either or both sides ensuring maximum powder penetration in the minimum possible space by creating a continually circulating powder cloud. Entrance and exit vestibules are fitted with air seals to prevent the escape of powder and the oversprayed material is recovered in a cyclone collector.

The "Compact Coater"

After many years of experience with hand-operated electrostatic spray units and automatic systems which were tailor-made to suit the individual requirements of the end user, it became apparent to the engineers at Volstatic that there was a need to design a safe, compact powder coating unit capable of satisfying the finishing requirements for the majority of their powder coating users. Their development effort has now resulted in the "Compact Coater" illustrated in figs 111 and 112.

Fig. 111. Coating chamber and control console of the "Compact Coater" (Courtesy Volstatic Ltd.)

The electrostatic spray coating chamber is a main feature of the "Coater", for the interior surface is constructed of a highly insulative dielectric material which repels powder from the chamber walls onto the components being passed through the chamber. This effect considerably improves the deposition efficiency of the powder coating. Spray booth assemblies are normally of metal construction and, to avoid random deposition of powder on the booth, it is necessary to site the emission point of the electrostatic spray gun at least 45 cm away from the walls. Due to the highly insulative material used in the construction of the "Compact Coater" it is possible to place the pistols close to the internal walls without reducing the deposition efficiency and this results in a considerable saving in the volume of the chamber.

The spray pistols may be adjusted for angle and position from outside the booth to achieve the most suitable spraying position. It has been found that a two pistol spray configuration is successful for most coating operations.

Fig. 112. Coating oven of the "Compact Coater" (Courtesy Volstatic Ltd.)

Oversprayed powder is collected from one end of the chamber after migrating longitudinally with a sweeping action of the powder on the walls of the chamber. Interlocking connections between the powder recovery unit and the powder delivery system ensure that emission of powder from the spray pistols does not occur if the recovery system is not in operation. An additional safeguard is the automatic current control which prevents sparking if the pistol comes into contact with an earthed object.

The powder delivery hopper is free standing to allow correct siting within the space available. Powder is completely fluidized and migrates to the spray pistols through self cleaning venturi ejectors.

A continuous loop conveyor of normal length 10·5 m is capable of handling the maximum load of components and the chain is enclosed in a cruciform cross section track to prevent excessive collection of dirt. The conveyor speed may be varied infinitely between 0·15 and 1·22 m per minute and the chain is protected against seizure by a torque limiting overload device.

A combination of radiant heat and natural air convection gives high overall efficiency to the in-line stoving oven with accurate control of oven temperature up to 350°C.

The entire operation is controlled from a single console which incorporates the safety devices included in the electrical circuits.

Although the "Coater" was designed as a complete package, it can be supplied in individual parts for integration with a customers existing installation.

The "Waggler" Unit

The spray guns may be fixed or attached to the "Waggler" device which causes the gun to move vertically about a horizontal axis by the operation of a motor driven offset cam.

Fig. 113. "Waggler" unit (Courtesy Volstatic Ltd.)

A reciprocator (which makes the gun traverse vertically) can also be included but this device is more prone to mechanical failure than the "Waggler" unit.

The "Autocoater" can also be fitted with a memory control system which will count the parts sprayed and switch off at the end of the run when the prescribed number have been sprayed.

A Carousel automatic unit is available for coating articles up to 3 ft. long by 1 ft. wide at high speed. Speeds of 30 ft. per minute have been achieved with this compact space saving unit.

The Field Mill Voltmeter

A useful piece of ancillary testing equipment supplied by Volstatic is the Field Mill Voltmeter.

This is a simple instrument for measuring the potential of a high voltage source without drawing a continuous constant current. It operates in the range 0–100 kV with either positive or negative polarity. The instrument is therefore suitable for measuring voltages from extremely high impedance

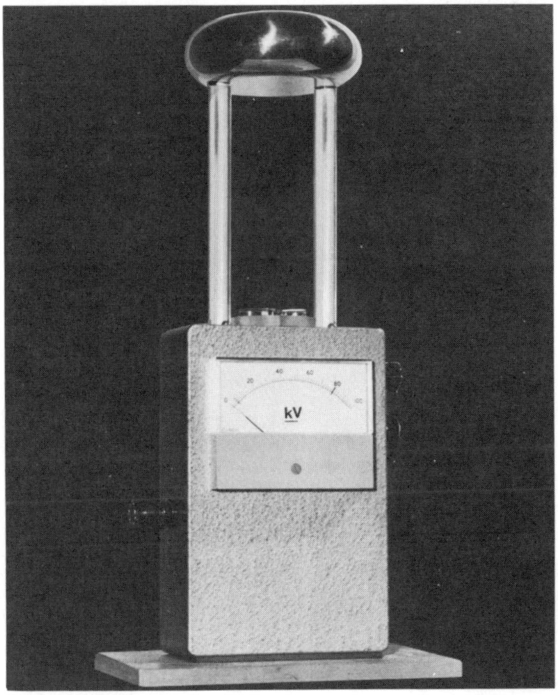

Fig. 114. The Field Mill voltmeter (Courtesy Volstatic Ltd.)

sources or where power outputs are of a low order and the current drain from a conventional voltmeter would be excessive and produce misleading readings.

The instrument has its best accuracy over the range 20–100 kV when the source is directly coupled to the high voltage terminal. When corona-coupling is employed the reading is not significantly in error, provided serious charging of the high-voltage insulator does not occur.

T.A.M./S.A.M.E.S./BINKS BULLOWS

The initial work with electrostatic powder application was carried out by S.A.M.E.S. in France during 1960 and S.A.M.E.S. powder coating units are now to be found in almost every country of the world.

In the United Kingdom the considerable experience of Binks Bullows as the leading manufacturer of spray finishing equipment has been combined with S.A.M.E.S. expertise to provide a comprehensive range of hand operated and fully automated powder spray units.

A complete technical service is offered and manual and automatic demonstration facilities are available.

Fully automated powder coating plants are specifically designed by Binks Bullows to suit the requirements of the end user and these may incorporate the most sophisticated modern methods of control.

The range of S.A.M.E.S. electrostatic spray guns includes the following:

Stajet 50, and Stajet 60 and Stajet 62

Three lightweight hand guns are offered in the S.A.M.E.S. range and they differ mainly in the design of the venturi feed section. A variable spray pattern can be obtained with each gun and this may be adjusted between the wide vortex cone pattern suitable for spraying flat surfaces, to a penetrating jet for coating the inside of box sections. A simple control at the back of the gun enables this variable pattern to be obtained.

The Stajet operates at voltages up to 90 kV. Voltages are supplied by a transistorized electrostatic generator and a safety device is incorporated to limit the current and discharge energy without interruption to the powder coating operation. The weight of all of these guns is 480 grams and the overall length 33 cm.

A maximum output of 25 kg of powder per hour is reported for the Stajet

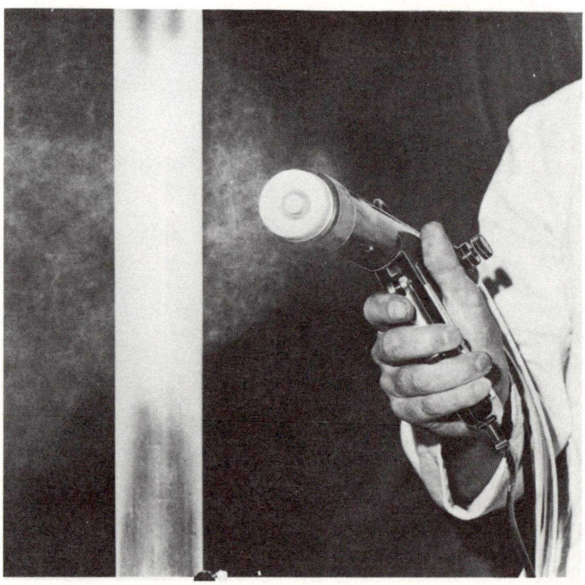

Fig. 115. Stajet 60 (Courtesy Binks Bullows Ltd.)

60. Extension barrels may be fitted to the hand guns to coat the interior of hollow articles or tubes.

The Stajet 62 has been designed to conform with B.A.S.E.E.F.A. and German P.T.B. standards.

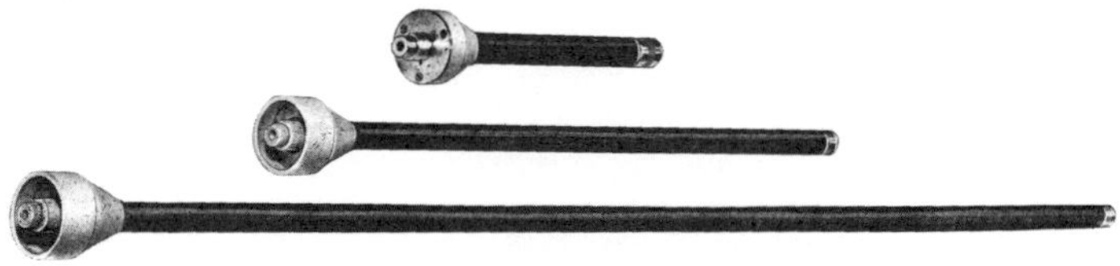

Fig. 116. Extension barrels for hand guns. (Courtesy Binks Bullows Ltd.)

The Megastajet

Megastajet SRV 83 is recommended for the automatic application of powder to complicated articles with an output of 50 kg per hour. Vortex nozzles in the gun ensure maximum electrification of the powder particles and the spraying head is fitted with an automatic unclogging device to prevent build up of packed powder. Weight of the SRV 83 is 4 kg.

Fig. 117. Megastajet SRV 83 (Courtesy Binks Bullows Ltd.)

Autostajet SRV 86

This is a light spray gun weighing only 700 grams which may be used in sets on horizontal or vertical reciprocators with each gun capable of delivering up to 25 kg of powder per hour. The gun uses the same nozzles as the Stajet series of guns with operating voltages up to 90 kV, and it requires no maintenance.

It is possible that the SRV 86 would be used also in automated powder systems operated by the computerized R.A.M.P. method.

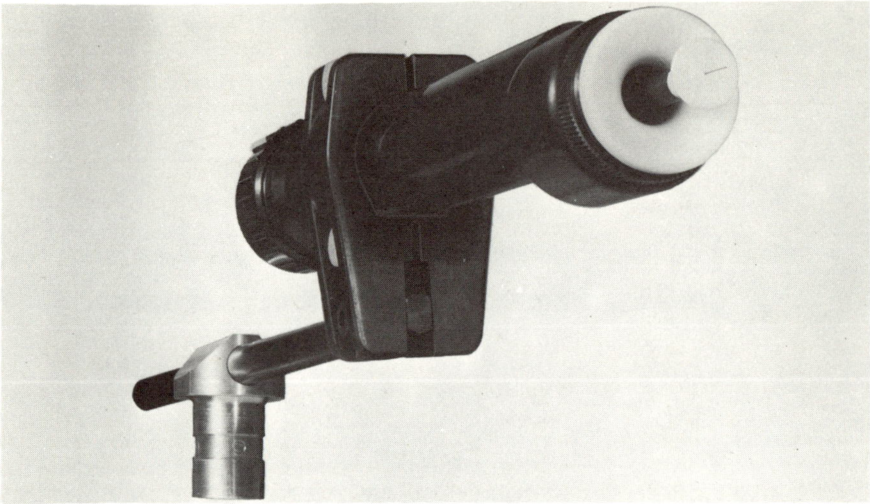

Fig. 118. Autostajet SRV 86 (Courtesy Binks Bullows Ltd.)

Computer Controlled Spray Application

R.A.M.P. is a computer controlled robot method simulating the actions of a skilled spray operator.

The initials refer to "random access multiprogramme" which means that the device can select at random any one of sixteen different painting programmes.

Separate programmes for each article to be coated are prepared by a skilled sprayer who operates the device manually to ensure complete and efficient coverage. Every movement of the spray gun is recorded on tape so that a faithful reproduction is obtained when the tape is replayed.

When all the items to be sprayed are programmed, they may be hung on the conveyor track in any order for the necessary coding is contained on the jig and a sensing device on the conveyor operates the control console.

The manipulator or spray unit consists of four main sections. A pedestal base which supports the cone unit and serves as a housing for the electrical connections. The cone unit, mounted on the pedestal base, contains the hydraulically driven mechanisms which produce the rotating and tilting motion of the horizontal arm. A stiff T section aluminium extrusion operates the horizontal arm which is pivoted at the centre. The wrist assembly is attached to the front tie block and contains two wrist action motors simulating pitch and yaw. The spray gun handle is attached to the pitch motor.

Control of the operation and the hydraulic power pump are each housed in separate free standing units.

R.A.M.P. was devised for automatic wet paint coating operations but the use of this equipment for electrostatic powder coating is already a reality that ensures complete automation without touch-up and very high transfer efficiency of powder.

Fig. 119. R.A.M.P. unit (Courtesy Binks Bullows Ltd.)

THE AIRSTATIC POWDER SPRAY SYSTEM

Mindon Engineering manufacture the Airstatic spray equipment which offers a completely new approach to powder spraying for in their equipment there is a total absence of high voltage generation.

Powder introduced to the spray gun is charged by friction causing the internal ionisation of each particle at an efficiency which depends upon the physical and chemical structure of the powder.

This process takes place in an ion generator incorporated in the gun itself. There is no corona discharge to produce a field between the gun and the article but simply a powder cloud of singly charged carriers. It is claimed that this method of charging the powder gives infinitely better penetration in cavities and corners, thereby minimizing the losses due to overspray and effectively overcoming the "Faraday Cage" effect. The coating of confined internal surfaces is also possible because the accidental contact of the gun nozzle with the sides of the component will not affect the emission of charged particles.

Total absence of electrical leads and high voltage generating equipment eliminates the capacitance build-up which can cause a spark discharge from the tip of the spray gun with the risk of a powder explosion.

The basic components of the Airstatic spray system are the fluidized bed powder hopper, control cabinet and the spray gun.

Powder is stored in a stainless steel fluidized bed which is aerated to give an even supply to the spray gun without surging. The powder hopper is easily removed for quick colour change.

A control cabinet incorporates the fluidized bed control valve with a pressure gauge, a gun pressure control valve with gauge and a powder injector assembly. The cabinet is fitted with quick release air hose connections,

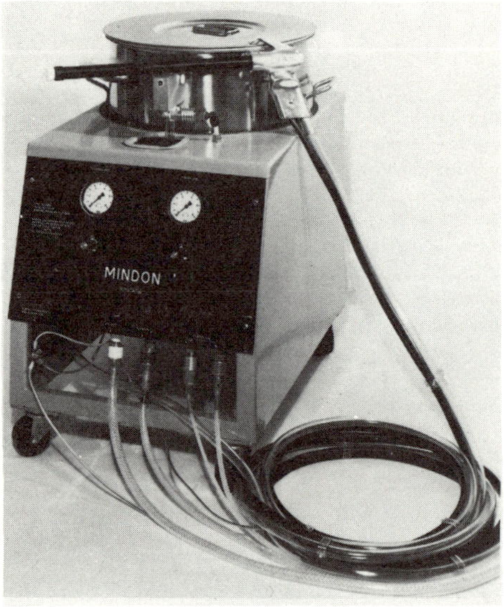

Fig. 120. The Airstatic unit (Courtesy Mindon Engineering Ltd.)

gun connections, airline and an earth lead. Casters are fitted for ease of movement.

The powder spray gun is extremely light in weight due to the absence of power cables and it possesses a small diameter barrel allowing easy access to intricate shaped articles. Various deflection nozzles are supplied to enable the spray pattern to be altered.

Maintenance is kept to a minimum by the simplicity of this equipment which utilizes only a compressed air supply.

There are three basic designs for the Airstatic spray gun. A laboratory "cup-gun" model is offered for small batch work and allows quick colour change for routine testing. Manual production models can handle up to 25 kg of powder per hour, and automatic arrangements incorporating one or several spray guns can be designed for fixed or reciprocating application methods to suit the requirements of the end-user.

NORDSON POWDER SPRAY SYSTEMS

Nordson are one of the leading manufacturers in the field of paint spray application and they have developed a range of powder spray equipment incorporating several unique features to comply with standards required by world authorities.

A complete technical service is offered together with test facilities.

Safety in operation is the prime feature of their entire range of application equipments with a patented electrostatic resistor cable and a resistor assembly in the gun to prevent arcing between the gun and the article. A further safety device is the non-electrical air switch in the hand gun which activates both powder flow and transfer of voltage and eliminates the possibility of sparks in the body of the gun. Current limiters are incorporated in the latest units to reduce the risk of arcing to earth.

When all the factors that can affect deposition are taken into consideration, Nordson claim an efficiency of deposition higher than that of any other powder application equipment. The spray gun incorporates a centrally placed external charging antenna which allows a uniform electrostatic charge to be given to all the powder particles.

A cone shaped deflector is claimed to improve the deposition efficiency by separating the powder from the powder/air mixture which imparts a low flight velocity to the powder particle.

Quick colour change, featured in the Nordson equipment, has been achieved by the provision of a disconnecting handle on the hand gun and quick change powder feeders on hand and automatic set-ups. One hopper has simply to be uncoupled from the venturi metering device and replaced by another hopper.

The venturi metering device is fed by powder from the bottom of the hopper which has been fluidized by an air diffuser which also breaks up the powder agglomerates before they enter the injector throat. Smooth powder flow up to 27 k per hour is regulated from the control panel.

The range of Nordson equipment includes:

Model NPE 1M Manual Spray Gun

The powder spray pattern is adjusted by moving the deflector forward to widen the pattern and retracting the deflector to give a narrow spray band. Operating voltages are variable between 30 and 90 kV with a maximum operating current of 200 μA. The handle of the gun is hinged so that the powder feed line may be removed and replaced without the use of tools, and a rapid colour change can be made by the insertion of a different powder feed line.

Model NPE F1 Feeder Power Unit

A combined powder hopper, electrostatic generator and control console is contained in the feeder power unit model NPE F1.

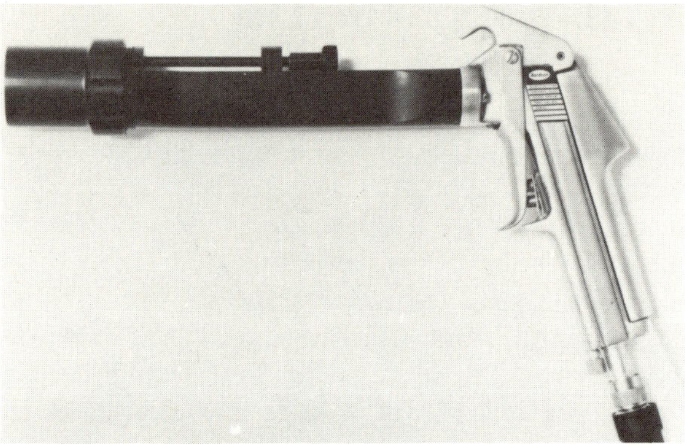

Fig. 121. The NPE 1M hand gun (Courtesy Nordson Ltd.)

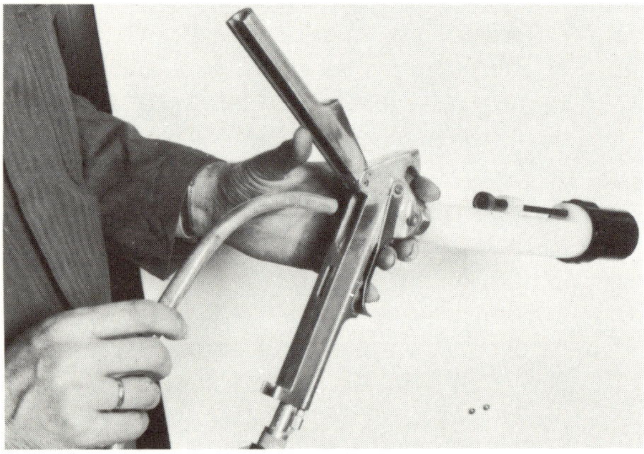

Fig. 122. Side view of NPE 1M showing quick disconnect feature (Courtesy Nordson Ltd.)

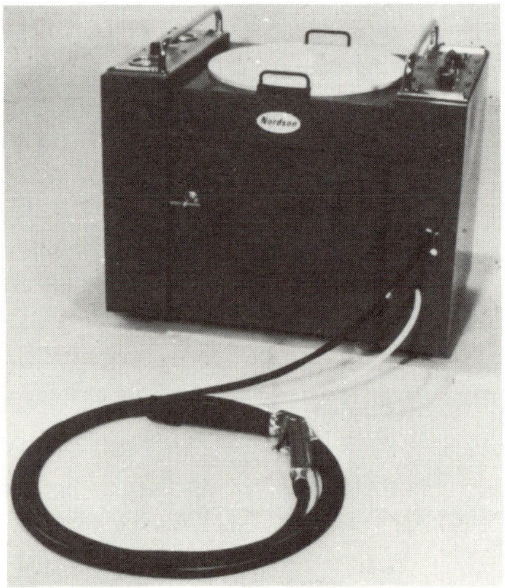

Fig. 123. The NPE F1 feeder unit with NPE 1M gun attached (Courtesy Nordson Ltd.)

Fig. 124. NPE F1 feeder with hopper being removed for colour change (Courtesy Nordson Ltd.)

The automatic spray guns are similar in design to the hand gun except that the switch control is operated from the main console.

Multigun automatic Nordson plants employ console cabinets with modular design of power units and pneumatic controls. The model NPE CC1 can supply power to four guns and is used in conjunction with the hopper/feeder model NPE F2 which can supply air and powder to four guns.

Designs are available for even larger automatic units and it is possible to connect multiple hopper systems together to increase output.

DEVILBISS POWDER SPRAY SYSTEMS

The Devilbiss Company are a major supplier of application equipment and they have worldwide representation for their sales and technical service.

Complete powder coating plants, designed to suit the requirements of the end-user are based upon trials carried out at the demonstration centre. When the plant is installed an efficient after-sales service is maintained.

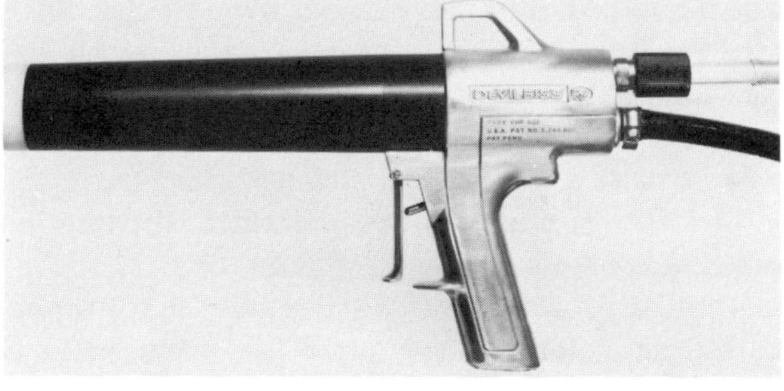

Fig. 125. EHP 502 hand gun (Courtesy Devilbiss Ltd.)

The principles in the design of the hand gun are also used in the fully automatic units and the main feature of the Devilbiss equipment is the simplicity of design which allows easy cleaning and it is claimed that the changeover from one colour to another can be effected in thirty seconds on the manual equipment, and in six seconds on the automatic equipment.

A microfeed powder cup can be attached to the hand gun for testing small quantities of powder, a unique feature which avoids the necessity to clean down feed hoppers when testing a new powder.

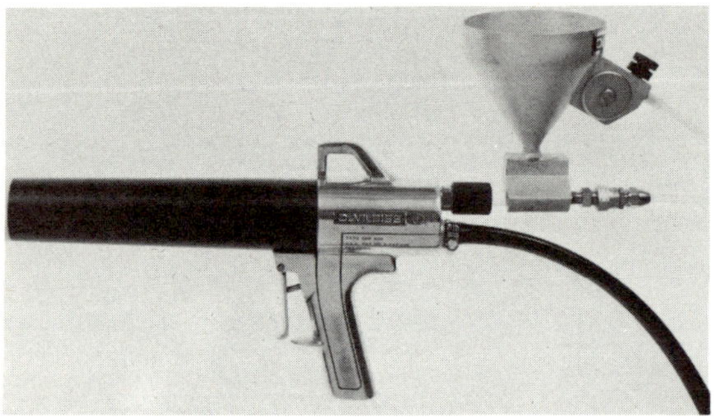

Fig. 126. EHP 502 with Microfeed cup (Courtesy Devilbiss Ltd.)

EHP 502 Electrostatic Gun

The gun contains a tubular powder passage through which the powder is conveyed to the vortex chamber in the nozzle. Vortex air is introduced to impart a swirling motion to the powder as it is carried past the high voltage electrode. The trigger action control operates the vortex air, powder feed and high voltage switch. Depressing the trigger half way produces a direct jet of powder for penetrating hollow sections and sharp corners while a full depression of the trigger gives a wide spray pattern suitable for the application of powder to flat sheet.

The EHP 462 electrostatic air cable carries the vortex air and high voltage to the gun. A separate hose supplies powder from the powder feed tank.

PFT 509 is the Devilbiss reference number for the powder feed tank, which operates on the fluidized bed principle to reduce agglomerations and packing of the powder particles. A powder pump unit is suspended in the fluidized cloud of powder and this withdraws powder when the trigger control is operated. The powder continues to be fluidized in the tank even when the hand gun is not in use.

Power supply unit EPS 514 has a negative output adjustable from 38 kV to 90 kV. The output voltage is regulated for variation in line and load current until a load current of 150 μA is reached. Thereafter the output current is maintained at 150 μA and the output voltage varies as required to

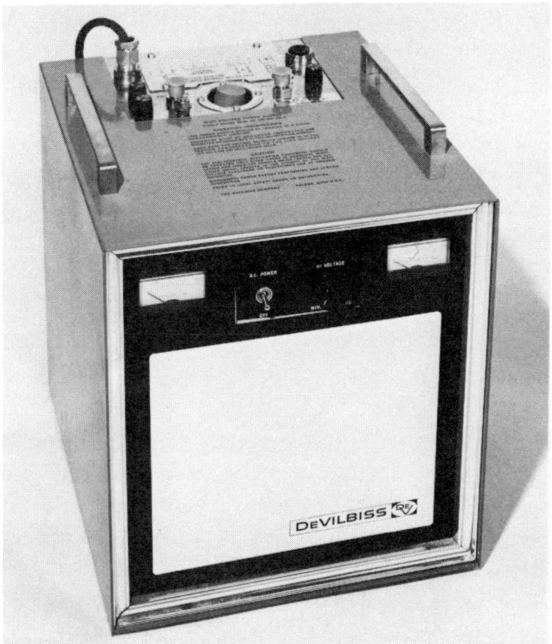

Fig. 127. Power supply unit EPS 514 (Courtesy Devilbiss Ltd.)

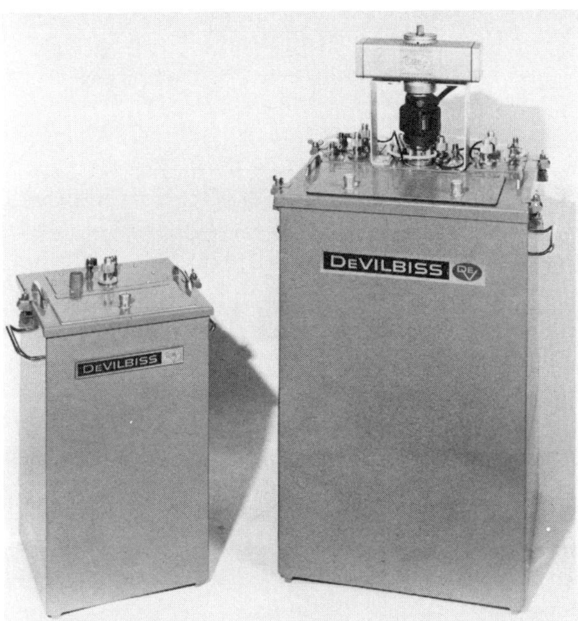

Fig. 128. Pneumatic control cabinet (Courtesy Devilbiss Ltd.)

limit the current to this value. If the load is removed the voltage will return to its preset maximum value.

Power and air supplies are interlocked so that an air pressure failure causes the high voltage to be turned off.

A pneumatic control cabinet allows variable control of both powder feed and vortex air.

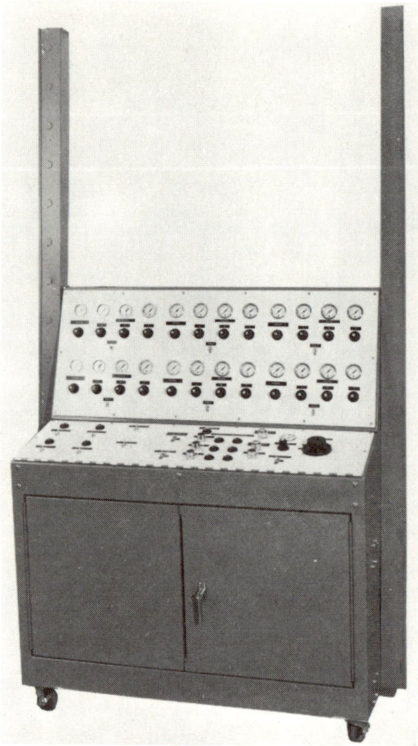

Fig. 129. Console operating up to six guns (Courtesy Devilbiss Ltd.)

The pneumatic control and power supply may be incorporated in a console operating up to six guns.

SPRAY FINISHING SYSTEMS

Spray Finishing Systems are a British Company with worldwide associations and they are able to offer a complete range of manual and automatic powder spray plants using the well known E.S.B. equipment developed in Germany by G. F. Vöhringer.

Test house facilities are available, and plant to suit the needs of the client can be designed when the trials have been completed.

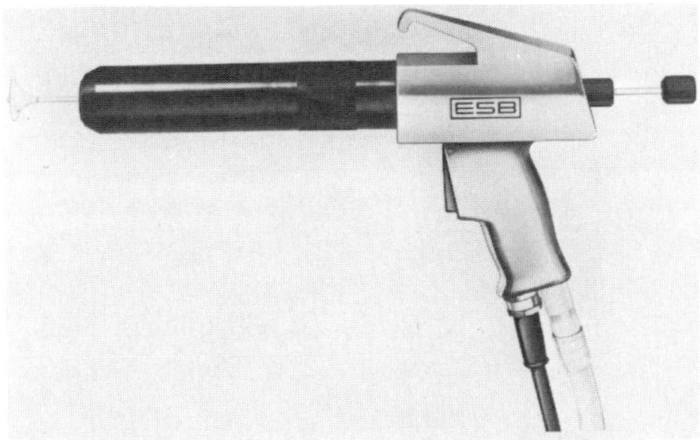

Fig. 130. Model 50493 hand gun (Courtesy Spray Finishing Systems Ltd.)

Features of the S.F.S.-E.S.B. equipment are the high charge transfer efficiency of the electrode in the spray nozzle with infinitely variable voltage adjustments to assist the coating internally of small hollow items. The generator module is designed to supply two guns separately with variable voltage, and the powder dosage to each gun is individually controllable.

Precise metering of powder to the gun is achieved by the pneumatically operated powder lock in the fluidized outlet of the powder hopper.

The spray cone angle is variable between 30° and 120° by simple adjustment to the round jet nozzle.

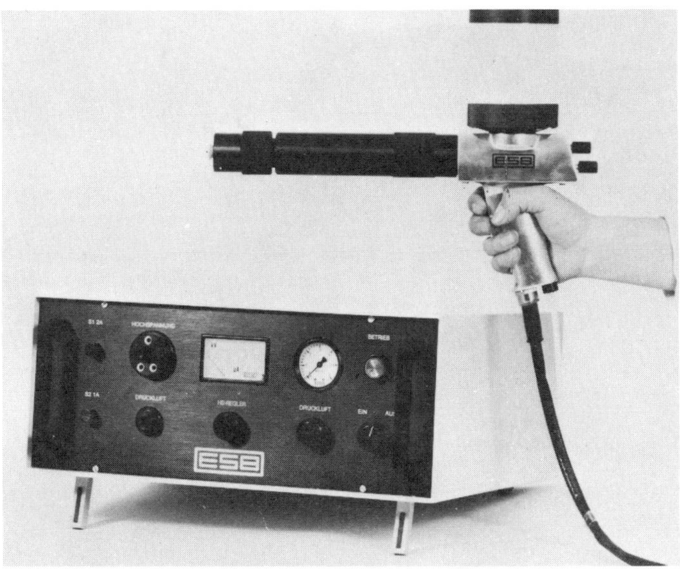

Fig. 131. Model 50458 laboratory powder unit (Courtesy Spray Finishing Systems Ltd.)

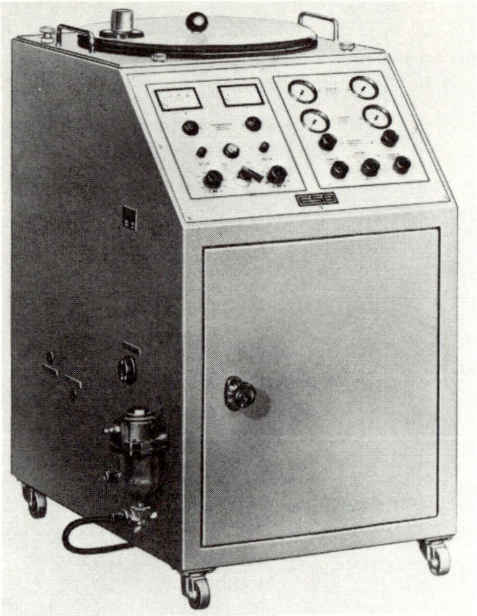

Fig. 132. Model 50411 ESB powder electrostatic unit (Courtesy Spray Finishing Systems Ltd.)

REFERENCES

1. R. P. Corbett. University of Southampton. Electrical Engineering Department
2. A. W. Bright. Prof. University of Southampton. Electrical Engineering Department.
3. A Golovoy. Ford Motor Co. Ltd. *Journal of Paint Technology* **45** No. 580.
4. R. C. Lever. Volstatic Ltd. Industrial Finishing. *Surface Coatings*. Dec. 1974.

11

Safety Procedures

SUMMARY

Considerable attention has been given to precautions which must be adopted by end-users of powder coatings, to ensure that powder explosions do not occur and that the health of operators is not impaired by exposing them to powder-laden atmospheres.

This action has emanated from a genuine desire by the powder coatings manufacturers to be strictly ethical in their association with the customer who purchases powder coatings. It is not motivated by any belief that powder coating is more hazardous than the wet paint processes, as it can be shown that danger is actually reduced when powder coating replaces most wet paint operations.

Manufacturers today, are concerned with the need to develop and maintain safe working procedures, so the emergence of a new technology offers a golden opportunity to provide safeguards from the outset. It also obviates the need to wait for danger to be made apparent when accidents occur and it eliminates being wise at a much later date when the health of workmen has been irreparably damaged.

In co-operation with Government bodies, the Powder Coatings Industry has assisted in the composition of Codes of Safe Practice[1] which are now leading to legislation affecting the design of spray guns and booths, the chemical composition of powder coatings, and safe working limits for concentrations of powder in air.

In this chapter, attention is first given to safety in the manufacture of powder coatings where little or nothing has been said to date about safe working procedures. Aspects which are discussed at length concern toxicity of components, the need for adequate ventilation and dust extraction, safety guards for machinery, explosion reliefs in fine grinding mills and, in addition to the obvious precautions, a general discussion on the important matter of "good house-keeping".

Protection for end-users is easier to formulate, for it is merely necessary to summarize legislation and safe working codes which are already established. This review will cover safety requirements for electrostatic spray equipment (drawn up by B.A.S.E.E.F.A.,) and testing organic powders to determine their safe working limits of concentration to prevent explosions and avoid toxicity hazards. Again, emphasis will be placed upon general conditions of hygiene.

It is a difficult and costly operation to introduce the essentials for safe working once plant has been installed, and prospective manufacturers or end users of powder coatings are urged to give the matter their full attention at the planning stage.

SAFETY IN POWDER MANUFACTURE

TOXICITY OF INGREDIENTS

Solid grade epoxy resins have been used in industry since the early 1950's, and they are accepted components of many internal container linings used in the food industry. The grades employed in the manufacture of powder coatings are similar to those included in the white list of the American Federal Drugs Act.

Certain low molecular weight diglycidyl ethers have been suspected to be carcinogenic, but these products are neither used wholly, nor partially, in powder coatings. The major suppliers of epoxy resins can offer authenticated evidence of the low level of toxicity of their products, based on extractability measurements and long-term feeding tests.

Polyester and acrylic resins, suitable for use in powder coatings, contain smaller quantities of low molecular weight fractions than the types used in wet paint systems, and these resins are unlikely to present a toxic hazard.

Pigments and extenders are mainly non-toxic, with the notable exception of the lead chromes and cadmium sulpho-selenides.

In Britain and the United States, the use of lead pigments is considered to be hazardous if the level of soluble lead contained is above a specified level, while cadmium pigments until recently have been considered acceptable. Swedish authorities, however, now look upon cadmium pigments as a greater health risk than lead chromes, so it seems that there is no consistent opinion about the relative hazards of the two classes of compound.

Safety legislation, devised to set standards for the Toy Industry, gives strict limits of tolerance for pigments based on most of the heavy metals.

Wherever possible the formulator will produce coloured coatings from organic pigments of low toxicity, but for certain colours, notably yellows, oranges and reds, it is only possible to formulate a finish with good flow and opacity by the use of lead or cadmium pigments. The choice of these pigments can, however, be confined to those grades which have acceptable levels of soluble pigment. An assurance on the toxic level of such pigments can be obtained from most of the reputable manufacturers, and the slightly higher cost of acceptable grades is a small price to pay for this guarantee.

Metallic powders and carbon blacks do not normally present toxic hazards.

The main pigment used in powder coatings is titanium dioxide and there are few chemicals which can be said to be less toxic than this widely used raw material.

Hardeners or curing agents are the materials which give most concern to the powder manufacturer and each new product introduced in the powder field has been subjected to a critical evaluation of toxicity hazard. Dicyandiamide is a fairly innocuous compound, but the additives used for accelerating the rate of cure can be irritating when powder dust is inhaled. The compounds used are considered non-toxic, however, at the low levels encountered in powder manufacture.

Acidic hardeners based on the dianhydrides are unpleasant to work with, for they affect the respiratory system, and, in fact, the German authorities consider these materials to be carcinogenic. They are now introducing legislation to prevent the use of these compounds as curing agents. The use of blocked isocyanates is a perfectly safe procedure, for they are stable up to temperatures of 160°C and volatile components released above this temperature are usually non-toxic.

Flow additives are mainly resinous and harmless.

Workmen employed in the paint industry can develop dermatitis in contact with the solvent-borne paints, but there is no evidence that this risk occurs in powder coatings. In fact, there are known cases where workmen who have suffered from dermatitis have had no recurrence of this complaint when transferred to powder coating manufacture.

It has been observed that dust produced in powder manufacture can block the pores of the skin and give rise to spots or pimples, but this is the only form of skin irritation evidenced to date.

DUST INHALATION

The problem of toxic hazard resulting from ingestion of powder coating ingredients is much less than the real risk to health which could result from continued inhalation of finely divided dust. Fibrous materials have been responsible for many cases of chronic invalidism and death due to different forms of silicosis in the mining industry and in the asbestos- and silica-using industries.

The particles used in powder coatings are not fibrous, but it would be unwise to be complacent about the effect of these roughly spherical particles on the walls of the lung.

Two precautions must be taken to protect workmen from this real health hazard.

(a) A dust mask must be provided.
(b) Adequate ventilation and efficient dust extraction must be provided at those production sites which are naturally dusty.

It is the writer's experience that workmen will not wear dust masks for long periods because they are often uncomfortable to breathe in when performing arduous tasks. In addition masks are invariably ill fitting due to poor design.

Lightweight masks must be provided for the workmen to wear for the relatively short periods when they are engaged in weighing out irritant hardeners at the premix stage, during the transfer of premix to the feed hopper and the weighing and packaging of finished product. The cost of a dust mask is minimal and there are numerous designs for permanent and disposable masks. Paper or fabric disposable masks are widely used by sprayers, sanders and polishers, and they can be worn with little discomfort.

They are not as effective in dealing with irritant dusts as are permanent masks, but they can be recommended for weighing and packaging of the finished products where dust is unlikely to be noxious and particle sizes are significantly higher than in the dusts evolved in premixing.

Each operator should be issued with his own permanent mask, together with a supply of replaceable filters, and experience has shown that diligent workmen will maintain their own personal protective masks at a high standard of cleanliness and efficiency.

The main criticism against the use of a dust mask is the sense of false security which can ärise from the belief that it offers complete freedom from risk of dust inhalation. At best it is merely a second line of defence to be used in conjunction with efficient dust extraction.

There is no real alternative to efficient dust extraction equipment to remove the risk of dust inhalation during powder coatings manufacture.

The installation of this equipment must be included at the design and layout

Fig. 133. Standard unit U.M.A. (Courtesy Dust Control Equipment Ltd.)

stage of all new powder coatings manufacturing plants, because it is difficult and considerably more costly to introduce at a later date.

Dust extraction withdraws large volumes of air from the working area, and if this air is exhausted to the atmosphere there will be very great heat losses which, in winter, are wasteful and costly.

It is therefore recommended, that an efficient dust filter is incorporated into a closed cycle system so that clean air can be returned to the factory.

In practice, it has been found that a single dust extraction unit is not sufficient to remove the dust produced at the several manufacturing sites, owing to the difficulty of balancing air extraction rates at varying distances from the fan.

It is possible, of course, to blank off those points where dust is only produced spasmodically, but continuous production lines tend to produce dust continuously at all sites.

A much better arrangement is to have individual dust extraction fans and filter units at each manufacturing site, and the size of the unit can be designed to provide the extraction efficiency required at that site.

An individual cyclone and filter unit could be provided, but this is a very costly and cumbersome method of supplying dust extraction.

The D.C.E. Unimaster range of dust control units can provide the desired extraction capability for every dusty operation encountered in powder coating manufacture.

Dust Control Equipment Ltd. manufacture five basic constructions which can be combined with a range of fan sizes, filtration areas and dust container capacities, to give a choice of more than five hundred dust control units.

The standard model UMA, shown in the sketch, is an integral unit complete with fan, easy access filter assembly, a triple inlet hopper and a dust container with a quick release sealer device.

A fan and filter unit only are incorporated in the hopper assembly UMAH, which has a flanged base for bolting directly to the dust containing hopper.

The sack tipping unit UMA includes the fan and filtration sections mounted upon a materials-discharge hopper.

A venting unit UMAV consists of the filter assembly only, and is specifically designed to ventilate process machinery and storage vessels, which may be operating under pressure.

The fifth basic design couples the UMAV venting unit with a hopper and dust container if it is necessary to site the unit away from the process equipment.

Principles of Operation

Contaminated dust-laden air is drawn through the inlet to the filter bed by the suction of the fan. Initially a separation takes place in the unit hopper, where a baffle deflects the heavier dust particles into the dust container below. Fine dust particles are carried up to the filter elements, where they are retained upon the outer surface of the filter fabric.

Fig. 134. Unimaster sack tipping unit (Courtesy Dust Control Equipment Ltd.)

Fig. 135. Filter elements being removed (Courtesy Dust Control Equipment Ltd.)

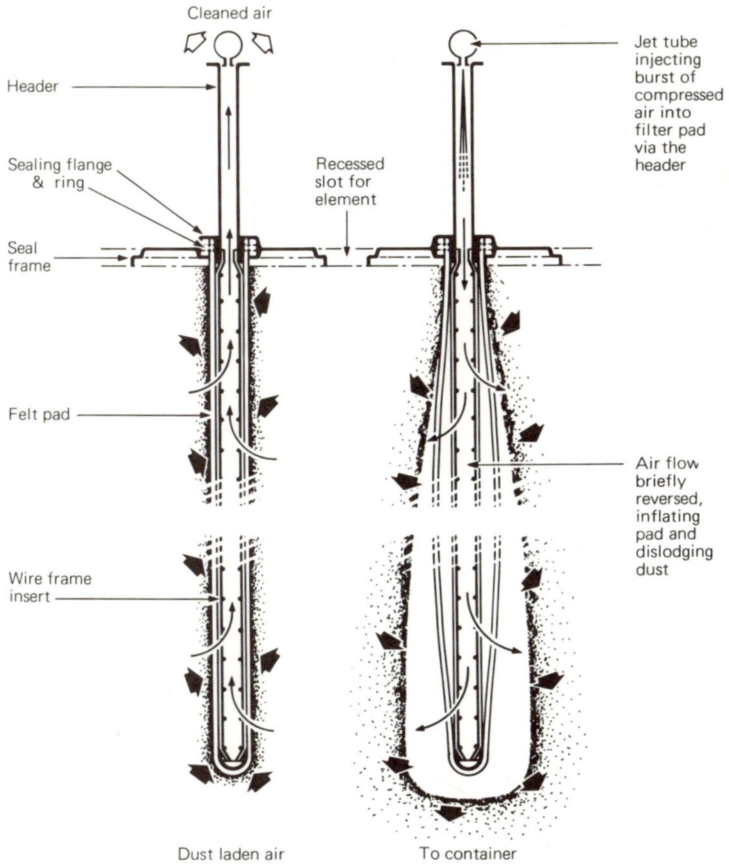

Fig. 136. Filtering and cleaning by reverse airjet (Courtesy Dust Control Equipment Ltd.)

Clean air passes through the filter fabric to the fan, and is then discharged. When the fan is shut down the cleaning mechanism is automatically activated, and the collected dust is dislodged from the filter elements and falls into the container. The cleaning mechanism comprises a shaker-bar, fitted with locators for the pads, and is driven by a motor mounted vertically on the side of the unit. The centre point on the bottom edge of each filter pad is positioned between two of the locators on the shaker-bar, so that during the cleaning cycle the pads are vibrated rapidly from side to side.

The filter assembly comprises a number of pad shaped fabric elements made up into a one piece bag. Each element has a flexible wire mesh insert which maintains the shape of the pads under air flow conditions. The whole assembly can be quickly withdrawn through the front of the unit. In the majority of applications a satin-weave pre-shrunk, cotton cloth is employed.

An alternative to the mechanically operated cleaning device is the reverse airjet cleaner, with an operating principle, which is more suitable for continuous dust extraction in conditions of heavy dust.

Advice on the correct equipment to suit a particular dust problem is best sought from a manufacturer producing a wide range of dust extractors.

Dust collection units used in powder coating manufacture should be fitted with an explosion relief panel.

Design of Extraction Sites

It will not be possible to purchase ready-made dust extraction inlets, so the design must be agreed at the planning stage of a new production line.

The construction and installation can then be undertaken by fabrication engineers when the equipment is placed in position.

The essentials to consider are:

(a) The volume of air to be extracted to correlate demand with the capability of the dust extraction unit.

(b) A simple damper should be installed close to the operating site, if several inlets to the dust collector are in use.

(c) The distance between the extraction point and the dust collection unit should be kept to a minimum to ensure high efficiency of extraction.

(d) Extraction points must not interfere with the work of the operator.

(e) Bends in the extraction pipe must be avoided where possible to prevent dust collection within the pipe which must be easy to dismantle for maintenance.

(f) To minimize the possibility of electrostatic charge build up, only metal fabrication to be used.

(g) A regular routine cleaning procedure must be adopted to maintain the efficiency of the dust collection unit. Removal of collected dust and an inspection of the filter bed should be made at the end of each working week.

(h) A filter cloth must be chosen which will remove all fine dust if the cleaned air is to be returned to the factory. In practice, it is unlikely that Factory Inspectors will permit cleaned air to be exhausted directly to the external atmosphere because a fault in the filter bed would not be immediately observed, and, without the manufacturer realizing it, dust could therefore be discharged into the neighbourhood.

Individual Dust Extraction Sites

WEIGHING STATION FOR PREMIXING SECTION

Scales used for weighing the premix constituents do not need to be larger than 100 kilo capacity, for the bulk of the premix charge will be made directly from preweighed sealed packages in which the pigments and resins are delivered.

A portable, lightweight canopy with adjustable height control would provide a versatile dust extraction point which would be suitable for the variety of container shapes and sizes used on the scales.

The portable canopy could be moved away from the scales and used as a temporary dust extractor over the open-ended drums, which may be charged for premixing (see Chapter 5).

A flexible wire-braced extraction duct should be employed with this mobile dust extractor.

It is important that spillages of dusty raw materials should be removed at once by the use of an industrial vacuum cleaner. (A broom employed for this purpose would produce more dust.)

PREMIXING SECTION (see chapter 5)

The majority of premixing machines will be permanently sited and fitted with a more permanent dust extraction device.

Dust extraction must be provided at the charging inlet to the premixer and at the discharge outlet. A damper device should be included at both extraction sites, since continuously operating dust extraction is not required and it would be convenient to connect both points to the same inlet of the dust collection unit.

A canopy is an efficient and simple device to instal at the charge inlet.

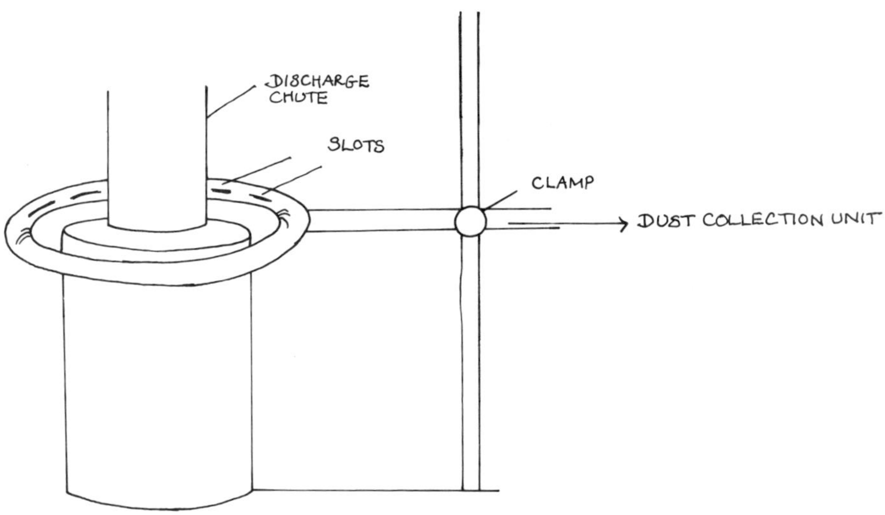

Fig. 137. Ring type extraction device

At the discharge point a ring-like extraction device would be less cumbersome than a booth, and quite efficient. This device could be adjustable and easy to remove for maintenance cleaning.

Between the charging and discharging operations the machine should perform the mixing operation without evolving dust, provided that the seals are replaced when they begin to wear.

COMPOUNDING SECTION (see Chaper 6)

The Z-blade mixer will require a permanent canopy which should be constructed of non-corrosive metal to resist the condensation of water evolved from the pigments.

Extruders require extraction points at the charge inlet to the feed hopper, and a possible extraction point at the extruder feed inlet, if the premix is

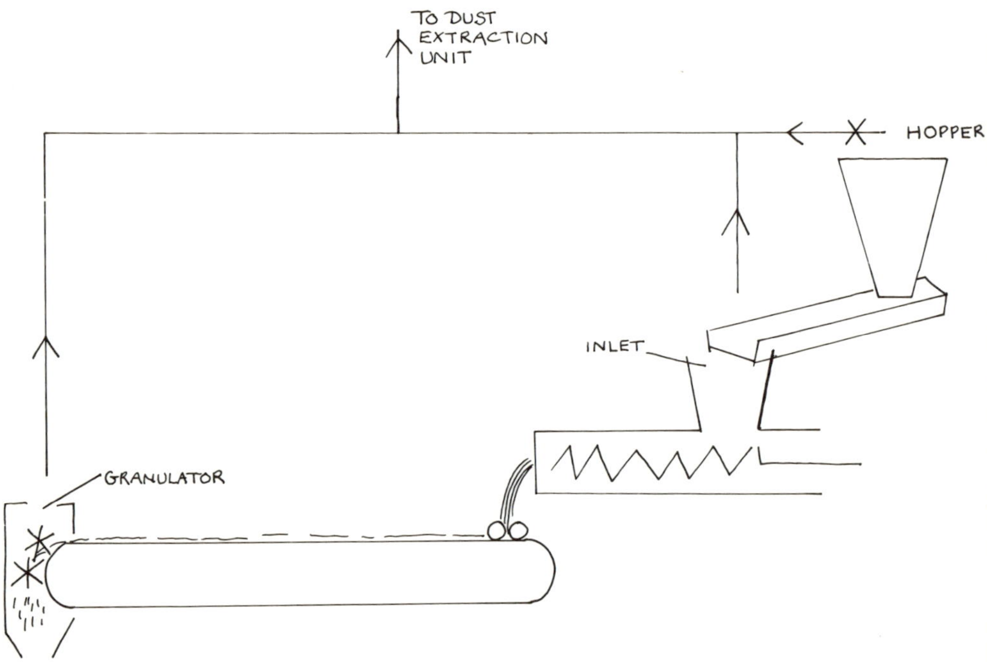

Fig. 138. Extraction points on extruder

transferred from the hopper by a vibratory feed trough. If a vibratory feed is used, the trough should be completely enclosed to minimize the dust problem. A third point of extraction is the granulating or flaking device sited at the end of the cooling band.

The type of dust extractor will be largely dictated by the geometry of the extruder and its ancillary equipment. A large output, continuously operating extruder can have the hopper designed to accept an automatic, totally enclosed feed from the premixer, so that dust extraction would be unnecessary at this point.

Charging the hopper with premix by hand, however, will be an intermittent and dusty process and an extraction point with a damper should be fitted.

At the extruder inlet and the granulator, continuous dust extraction will be required and dampers need not be fitted.

The extrudate discharge point is not dusty, but some water vapour and fume can be evolved. With the larger extruders it may be necessary to instal some extraction at the discharge point.

GRANULATING MACHINERY (see Chapter 7)

These machines are very dusty in operation and should be treated in the same manner as the premixing equipment. Throat fed granulators are best fitted for extraction by connecting the extractor pipe directly into the body of the charging section.

A ring type extractor should be installed at the discharge point.

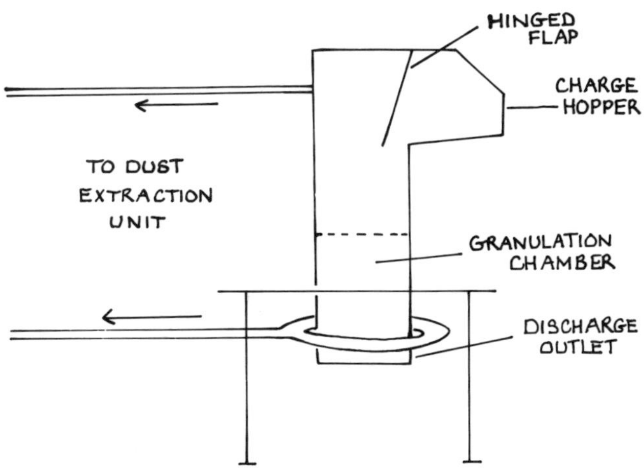

Fig. 139. Granulator and extraction sites

FINE GRINDING MACHINERY (see Chapter 7)

Grinding mills with high volume throughput rates have effective dust collection chambers, so that, during the operating cycle, in comminuting from granule to fine powder, there is little likelihood of dust seeping from the mill or collection filters.

Dust is produced at the charging point where the granulated feed is placed in the hopper, and at the discharge outlet for the finely ground powder.

A canopy extraction hood should be sited over the feed hopper allowing sufficient head-room for the operator to charge manually if required.

An extraction point, at the discharging rotary valve beneath the fine powder collection hopper, is unnecessary if the collection unit is maintained under reduced pressure. Fine dust discharged by the rotary valve is sucked back into the collection hopper, while the heavier non-dusty particles fall into a container beneath the valve. This feature is present in the blow-ring operated collection unit described in Chapter 7, and it is the only really dust-free continuously operating system.

Many fine grinding machines have high internal pressure in their discharge hoppers and it is therefore necessary to discharge into a bag tied tightly to the discharge outlet. This has the obvious disadvantage that the bags have to be changed frequently, and it is difficult to assess when the bag is full enough to require changing.

If the rotary valve operating under high internal air pressure is allowed to discharge into a collector, without providing an air tight seal between the valve and collector, it would be essential to provide dust extraction at this point. However, the considerable amount of dust discharged would be a contributory factor to reduced yields of the finished product, and it would be wise to instal the more costly system operating under reduced pressure, since this has no adverse effect upon the yield.

To summarize, dust extraction at the fine grinding stage is not only costly to instal and operate, it will also reduce the yield of finished product. A system which reduces the need for dust extraction will prove to be the most economic proposition in the long term.

SIEVING AND PACKING OPERATION (see Chapter 8)

While most sieving machines are relatively dust-free in operation, providing the seals are maintained on a regular servicing rota, there is always the dust problem at charging and discharging.

An adjustable ring extraction point is the best method for controlling dust during charging.

At the discharge outlet, the dust problem can be avoided if the powder is fed into a polythene liner bag inside a cardboard carton, with the polythene liner tied to the discharge chute; the bag is then changed when it is filled to the correct weight. This is a high labour-intensive operation as the sieving machine is not capable of continuous operation. However, the method is preferable to the high losses of product which could occur if efficient dust extraction were provided.

SAFETY GUARDS

These are normally considered to be devices fitted to machinery to protect the operator from accidental injury. They can also serve as additional protection for the machinery to prevent damage during operation. Protective clothing is an additional aspect of safety which it is right to include in this section as a safeguard.

Machinery Guards

PREMIXING

Premixing machinery must only be operated with the following safeguards: The hinged lid in the charging inlet must be fitted with a trip-switch, so that the machine cannot operate if the lid is open. A protective grid between the lid and the body of the premixing machine will provide an additional guard against the operator reaching inside the machine while it is still in operation.

Since the premixing machine needs to be running while the premix is being discharged, an essential safeguard is a wire mesh grill, bolted to the discharge chute, to prevent the operator reaching up into the body of the machine.

COMPOUNDING BY Z-BLADE MIXER

Z-blade mixers must also be fitted with a robust metallic grid, with safety cut-out switch to break the power circuit when the grid is open. While a hinged lid should also be provided, it is recommended that the metallic grid should operate the safety cut-out, so that materials may be charged when the machine is operating.

The finished product is discharged from thè Z-blade by tilting the machine and pouring out the liquid. To avoid a surge of molten material, an "inching" control button should be included in the tilting controls.

COMPOUNDING BY EXTRUSION

Some of the extruders described in Chapter 6 are fed by mechanically agitated hoppers, and a safety grid must be fitted to the charging point of the hopper to prevent the operator reaching inside while the agitator is working.

This type of feed hopper is invariably seated directly on the extruder inlet, and it is not necessary to provide an extra safeguard at this point.

A vibratory hopper does not require the metallic safety grid at the hopper inlet, because there are no moving parts to cause injury. It will however be necessary to provide a metallic grid at the extruder inlet to prevent access by the operator.

COOLING ROLLS

A machine guard will be necessary to prevent operators getting their hands trapped in the cooling rolls. Foot operated switches should be located adjacent to the cooling rolls, so that the machine can be quickly stopped in an emergency.

GRANULATORS

It is not possible to place a protective grid in the granulator inlet, for this would prevent the feed entering the granulating chamber. The manufacturers of granulating machinery provide entry throats which are sufficiently deep to prevent the operator reaching the rotary cutting knives. The discharge outlet can be protected by the provision of a safety grid. A safety cut-out should also operate when the entry throat is separated from the granulating chamber during the cleaning process.

FINE GRINDING MACHINES

Where the feed hoppers are discharged by a motor-driven worm-screw-feed, it will be necessary to incorporate a protective metal grid large enough to allow unimpaired passage of granules, yet small enough to prevent access by the operator. Power operated rotary valves at the discharge of the collection hopper will require a metal grid safety guard.

SAFETY GUARDS FOR PREVENTING DAMAGE TO MACHINERY

If metallic objects such as tools, nuts and bolts are accidentally dropped into powder producing machinery while it is operating, a considerable amount of damage can result. The metallic grids already recommended to protect operators from accidental injury will also prevent large metallic objects from entering the machine. Additional safeguards to protect the machinery, which are also capable of further reducing the risk of damage to the machinery, include: magnetic separators attached to the metallic grid to retain nuts and bolts inadvertently charged to feed hoppers of extruders and fine grinding machines, and an overload cut-out to stop the machine before excessive amounts of damage are caused. The latter is particularly important to protect the barrel and screw of the costly extruder.

All machinery should be earthed to prevent electrostatic build-up.

Safety Clothing

Workmen engaged in powder coating manufacture should be supplied with safety footwear with conductive soles to prevent excessive build-up of electrostatic charge.[4] The most important item of safety wear is a shatter-proof eye shield to prevent particles of granulated powders damaging the eyes.

EXPLOSION RELIEFS

Dust collection units and fine grinding collection units are two pieces of equipment which should be fitted with an explosion relief. In a later section of this chapter, it will be stated that dust explosions can occur when the concentration of dust in air lies between certain limits, and when a source of energy sufficient to initiate the explosion is also present. These two conditions could be realized in powder collection units, where the source of energy might be derived from electrostatic charge build-up resulting from friction between powder particles. In practice these two well-defined levels of powder concentration and energy seldom occur, but provision must be made for the chance circumstances which could result in an explosion.

The explosion relief must be designed to channel away from the working area the pressure rise within the collection unit. This can be achieved with a bursting disc capable of withstanding no more than 4–6 p.s.i. internal pressure located in robust metallic trunking to act as the relief passage. Preferably, the relief should be directed through the roof of the building, or through an outside wall at a height where no damage could be caused to people passing by. The ducting should be no longer than ten feet measured from the explosion relief point to the outside wall or roof, and there must be no bends in the ducting to impede the passage of the released pressure.

A regular routine inspection of this important safeguard must be carried out to ensure that it is maintained at maximum efficiency.

Guidance in this matter will be given by the manufacturers of the dust collection units who often provide explosion reliefs as a standard feature of their equipment; these safeguards must, above all, satisfy the stringent requirements of the Factory Inspector.

NOISE IN POWDER MANUFACTURE

Fine grinding machines are noisy in operation and even a small hammer mill with an output of twenty to thirty kilos per hour can produce a noise level which is uncomfortable to the operator.

Noise is generated principally by the mill operating at rotor speeds in the region of 5000–8000 r.p.m., and the noise level reaches its highest intensity just before the granulated material is fed into the grinding chamber. A mill which is not under load can produce noise in excess of 110 decibels and this is well above the maximum acceptable noise level for factory working. Continuous high noise levels can adversely affect the hearing of the operators and their general mental health and well-being also suffers as a direct result.

Precautions must be taken considerably to reduce working noise and noise measurement assessment should be made when the mill is not being fed since a marked noise reduction of twenty to thirty decibels does occur when the feed enters the grinding chamber.

A small grinding machine is often more noisy than a large mill, and hammer mills are significantly noisier than pin disc mills.

Small mills generally operate at higher rotor speeds than larger mills and vibrate more readily due to their lightweight construction. The noise output from a small mill can be reduced considerably if the mill is totally enclosed within a booth constructed from a metal frame with handing curtains of heavy duty, flexible, polyurethane foam. A sliding door provides access for the operator to stop and start the mill and to fill the feed hopper. This is a most effective device and the noise level can be reduced to less than 20% of the unshielded working noise.

Larger machines are not easy to enclose totally and it is often necessary for the operator to be stationed close to the mill. In this event the workman must be protected by the provision of approved ear-muffs. These are not uncomfortable to wear and are more readily accepted by operators than is the wearing of dust masks.

It is possible to reduce noise levels of larger machines but each make and model of fine grinding machines will require individual attention.

The points to consider for reducing noise output are:

1. Operating mill rotors at speeds of no more than 4,500 r.p.m.,
2. Incorporating an air filter at the air inlet to the mill,
3. Fitting the air exit from the mill with a silencer or damping box.
4. Completely enclosing the point where powder is collected in vertical socks or filter bags with a foam-lined metal cylinder (allowing, of course, an exit through the roof to provide an explosion relief.)
5. Mounting the mill on cushions to reduce the vibration between the mill and the supporting frame.
6. Enclosing, if possible, the mill itself in a sound damping compartment.

Attention to these aspects of mill installation should easily reduce the noise level below 70 decibels.

ADDITIONAL PRECAUTIONS

These are a miscellaneous collection of "do's and dont's" which are often applied in industrial processes under the general description of "good housekeeping".

The absence of inflammable solvent from the manufacturing procedures greatly reduces the fire and explosion risks. However, operators working in dust laden atmospheres should be prevented from smoking and notices should be displayed to this affect.

Dust should never be collected by sweeping with a broom, and a good quality industrial vacuum cleaner must be used regularly to remove dust spillages and build-ups. Extension pipes for the vacuum cleaner are needed to clean high level window sills and roofing girders. The building used for powder coating manufacture should be designed to minimize dust collection points and to facilitate routine daily cleaning.

Workmen should be warned against the use of compressed air lines for cleaning machinery and, under no circumstances, should they be allowed to clean themselves down by use of a compressed air hose. High pressure air can cause serious damage to eyes, nasal passages etc.

Dust, collecting on the hands and arms of the operators, is easily removed by rinsing with tepid water. A shower should be provided for the use of workmen at the end of the work shift.

The general conditions of hygiene are detailed in the booklet "Safety, Health and Welfare Booklet No. 5", issued by the Factory Inspectorate.

SAFETY IN POWDER APPLICATION

TOXICITY OF POWDER COATINGS

In the section relating to powder coating manufacture it was important to consider the toxicity of the individual components and the effect they could have upon the workmen especially at the premixing stage. The powder, ready for application, is invariably supplied in the form where each particle is a homogeneous blend of components which cannot be separated during the application, collection and re-application of the powder coating.

Curing agents employed in powder coatings are often irritating to eyes and throat before they are hot-melt blended during the extrusion stage of manufacture. In the finished powder coating it should not be possible to detect the presence of these substances, and it is unlikely that operators will suffer any discomfort from a correctly compounded powder coating. There are however a few powder coatings which are still manufactured by the earlier technique of dry blending the hardener system, and these could cause some irritation to operators. An assurance should always be sought from the supplier that the hardener system is incorporated into their product by hot-melt blending.

Generally, powder coatings are neither toxic nor are they unpleasant to use at the application stage. Unlike wet paint systems they contain no organic solvents and therefore release no volatile vapours to affect the operator. They are described as "nuisance" dusts in the publication "Threshold Limit Values, Technical Note 2/71", issued by the Factory Inspectorate.

The new Industrial Health and Safety at Work Act requires the supplier to state whether any health hazard exists in the use of his products, and the existing end-user of powder coatings will be aware that the assurances already given show powder coatings to be almost entirely free from toxic hazard.

There are no known cases of dermatitis caused by contact with powder coatings and, in fact, operators who have suffered with this complaint when applying wet paint, have experienced no discomfort when they have changed over to hand operated powder coating guns.

DUST INHALATION

The problem, and the precautions to be taken against it, have been discussed at some length in the early part of the chapter. To make absolutely sure that the health of the operators is not damaged by exposure to dusty atmospheres, the following precautions must be taken:

(a) Dust masks must be provided, and the operator must be told that these masks are only a partial protection against dust inhalation.

The use of a dust mask is an additional safety measure and must never be employed as an alternative to the efficient dust extraction of a well ventilated, powder spray booth.

(b) Dust extraction is an essential feature of all reputable powder spray booths and collection systems. The operator should be stationed outside the booth and the exhaust ventilation at the booth inlet should not be less that 0·4 m/sec (75 ft/min).

(c) Operators who are known to suffer with respiratory complaints should not be employed as powder applicators.

THE DUST EXPLOSION HAZARD

The basic principles of safe working practices are detailed in the Department of Employment's Health and Safety at Work publication—Booklet 22. "Dust Explosions in Factories" while a comprehensive study of the explosibility of dusts used in the Plastics Industry is given in the U.S. Bureau of Mines Report 5971. The figures given in the American report are based upon practical tests; similar procedures are used at the Fire Research Station, Boreham Wood. These tests give information about the ignition temperature of dust clouds and layers, the minimum explosive concentration, the minimum energy required for ignition, the explosive pressure generated, the rates of pressure rise at different concentrations and the limiting oxygen content of the atmosphere which would prevent ignition by heat or spark. West German authorities have also devised practical testing methods to determine these explosive properties, but they have found the determination of the minimum explosive concentration (sometimes referred to as the lower explosive limit) to be difficult to determine by practical methods, and a method developed by Selle and Zehr[5] is used to calculate the minimum explosive concentration.

Other factors which have been assessed are particle size and particle shape.

There are certain conditions which can give rise to the possibility of an explosion.

(a) The concentration of powder in air must be above a certain level, defined as the minimum explosive concentration.

(b) A source of energy must be present of a sufficient strength to initiate the explosion.

(c) The dust could come into contact with a hot surface whose temperature is above the auto-ignition temperature of the powder.

Minimum Explosive Concentration

Tests at the Fire Research Station have shown the minimum explosive concentration for epoxy based powder coatings to bc in the region of 20–30 grams per cubic metre, while the West German method gives a corresponding figure in the region of 50 grams per cubic metre. There is an upper limit to the concentration of powder in air, referred to as the maximum explosive concentration, above which an explosion will not occur, This figure is not determined in practice but is probably as high as 2 kilograms per cubic metre.

In the absence of a determined figure for the minimum explosive concentration, the spray booth should be designed so that a concentration of 10 grams per cubic metre is never exceeded. There has been some criticism of this maximum allowable concentration, since test results have been reported for powder coatings with minimum explosive concentrations below this arbitrary figure.

In the United Kingdom the average concentration of powder in the spray booth is determined from the rate of powder emitted by the guns in relation to the volume of air extracted by the ventilation system. This differs from the West German interpretation, where the total volume of powder delivered, minus that required to coat the articles, is expressed as a concentration in the air passing through the booth.

The explosibility of powders is affected by the particle size, but tests have shown that below 75 microns, the ignition temperature, minimum explosive concentration and the explosion pressure are relatively unaffected.[(6)] This same report observes that irregular shaped particles present a greater explosion hazard than spherical particles.

At the present time there is no evidence that polyester and acrylic powders are a greater explosion risk than the epoxies.

Aluminium-containing epoxy powder coatings have minimum explosive concentrations comparable with those of standard epoxy powders, but they give higher figures for the rate of pressure rise. Conversely, the inclusion of inorganic pigments and extenders will reduce the rate of pressure rise, although the lower levels of explosivity are not noticeably altered.

To ensure that the powder coating plant is perfectly safe during operation, the working concentration of powder in air should not exceed 50% of the determined minimum explosive concentration, and this rule must be applied to both manual and automatic spray booths.

Once again it must be emphasized that the limiting concentration of powder in air should be set at 10 grams per cubic metre and this limit should be used in the absence of determined figures for minimum explosive concentration.

It is also important to provide a smooth surface for the booth interior to prevent build-up of powder deposits, which could give rise to explosive conditions if disturbed. A feature of dust explosions is the chain reaction which can result from the disturbance of dusts by a minor primary explosion, leading to a more severe secondary explosion.

The following test centres are available in Europe for the assessment of powder safety limits.

The Fire Research Station, Boreham Wood, Hertfordshire, England.

Laboratoire du Centre d'Etudes et de Recherches des Charbonnages de France, CERCHAR—Industrie, Verneuil-en-Halatte, France.
Bundesanstalt für Material-prüfung, B.A.M., Berlin, West Germany.

Explosive Dust Clouds

Dust clouds in air require 50–100 times the energy necessary to ignite the same concentration of solvent vapour in air, so that powder coating processes are intrinsically safer than the majority of electrostatically applied wet paint processes.

It is not usual to specify a minimum ignition energy for powder/air mixtures above the minimum explosive concentration, for it is difficult to assess a finite threshold value for this energy level. The investigation carried out by the U.S. Bureau of Mines had indicated a minimum ignition energy for epoxy based powders in the region of 15–35 millijoules. In the section describing electrostatic spray equipment, it was mentioned that safety cut-out switches were incorporated into the circuits to limit the current employed to a safe level of approximately 100 microamperes. (1 millijoule = 1000 micro amps per 1 second = 100 microamps for 10 secs, 15–35 millijoules = 100 microamps for 150 to 350 secs).

A standard testing procedure for hand operated electrostatic powder coating spray guns has been prepared by B.A.S.E.E.F.A.[3], in conjunction with equipment suppliers, H.M. Factory Inspectorate and the Fire Research Station. This Certification Standard is intended to assist manufacturers by providing information and safety limits on the design, construction and operation of electrostatic spray guns, and also on the method of testing such guns. Additionally, conditions for the safe use of the equipment with flammable paints or powders are given, over and above those which would normally apply in a paint or powder spraying area. The research and testing work was carried out at the Safety in Mines Research Establishment and the Fire Research Station. A Certification of compliance with the B.A.S.E.E.F.A. standard does not imply that the certified equipment is universally acceptable in all industries, or in other countries where more stringent standards may be required.

However, it is a first and most welcome step towards the attainment of standards of safe working with electrostatic spray apparatus.

The testing procedure for hand operated electrostatic powder spray guns pays particular attention to the basic requirements of construction.

Impact and drop tests measure the resistance of the casing to accidental damage, and there is an assessment of the degree of protection afforded by the casing against damage by solvent, where plastics are used in construction, and the protection afforded by the casing against moisture and dust.

An impact test is applied to the parts of the spray gun which are considered to be weakest, using a hardened steel hemisphere of 25 mm diameter, weighing 1 kilogram, falling from a height of 0·7 metres. The fully assembled sample is made to rest in a stable position on hardwood blocks on a concrete floor. The apparatus must remain safe after this test.

A drop test is carried out upon two fully assembled samples by dropping from a height of 2 metres on to a concrete floor. The spray gun may be dropped from any position and the number of drops is left to the discretion of the testing authority.

Once again, compliance with the safety standard is only met if the apparatus is safe to use following this test.

Spray guns with plastic casings must have a minimum area of 10 sq. cm of metal under the operators hand, on each side of the handle.

All cables attached to the spray gun, with the exception of cables containing only intrinsically safe circuits, are subjected to a Cable Pull Test of 150 Newtons for 60 seconds. The energy of the power supply must be kept to a level where a fault in the cable does not ignite itself or powder-deposits.

In the event of damage to the cable, the power supply should be designed to minimize the danger of electric shock. Where the circuits in the cable are not intrinsically safe, a metallic screen must be used to protect the high voltage cable, and this in turn is to be protected by an insulated sheath. The spray gun and the associated high voltage cable must withstand an insulation test of 1·5 times their maximum working voltage.

Other points of electrical safety which are examined include; the provision of a reliable connection between the metal parts of the spray gun and the earthing terminal in the power supply, and the preferred method of connection is via the metallic screen used to protect the cable; the trigger of the spray gun must be spring loaded towards the "off" position and it must be a high voltage control; any current-limiting resistor used in the spray gun must be insulated and protected against the possibility of a short circuit, and it must be adequately rated against a short circuit between the gun-nozzle and earth.

Auto-ignition Temperature

Perhaps the most important test is the operation of the spray gun in a flammable atmosphere.

A test apparatus is prepared as shown in fig. 140 and the spray gun is set to give the maximum possible H.T. voltage provided by its generator. For powder spray guns the test chamber is filled with the following gas mixture.

$$5{\cdot}7 \pm 0{\cdot}3\% \text{ Methane}$$
$$13{\cdot}2 \pm 0{\cdot}3\% \text{ Oxygen}$$
$$81{\cdot}1 \pm 1{\cdot}6\% \text{ Nitrogen}$$

The rotating electrode system is operated for at least 4 minutes and the gas mixture must not be ignited for the gun to pass this test.

The B.A.S.E.E.F.A. Standard also makes the following recommendations for hand operated electrostatic powder spray guns, which should be conveyed to all operators during their training programme.

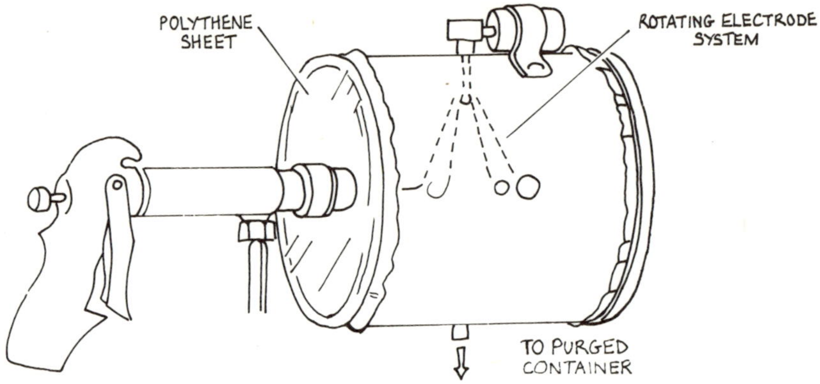

Fig. 140. Ignition chamber with gun in position

Electrostatic spraying should be carried out in a special booth designed for that purpose. This booth must be a fire resistant structure provided with efficient exhaust ventilation, and located close to an external wall to minimize the length of the extraction ducting. The internal surface of the booth must be smooth and free from dust collection points. It is important that all fittings within 3 metres of the spraying area should be earthed to prevent the excessive build-up of electrostatic charge.

The floor must also be electrically conductive, and in this respect bare concrete floors are usually found to be acceptable. Coverings which are electrically insulative must not be used on the floor of the spraying area, for the operator must be earthed at all times. Where a powder recovery system

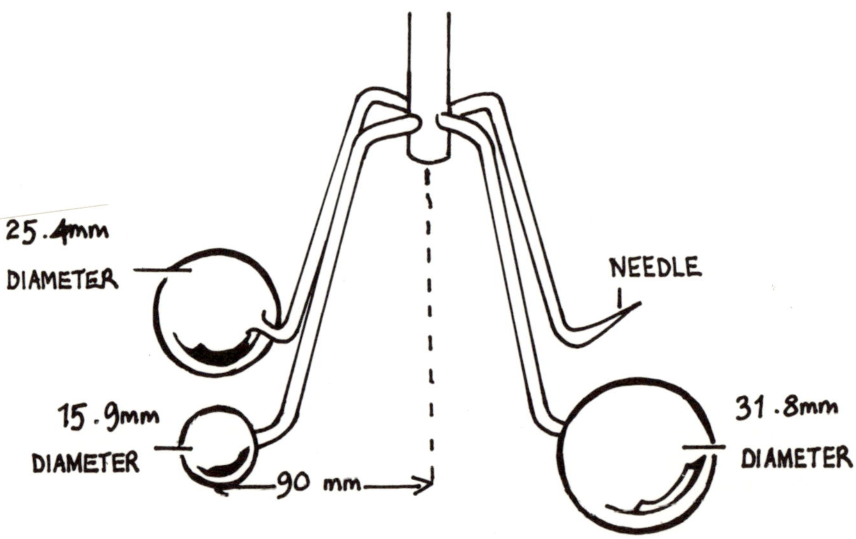

Fig. 141. Rotating electrode system

is incorporated, adequate precautions must be taken against the risk of fire and explosion. These are matters which have previously been discussed in this section but a more comprehensive account is given in the Department of Employment's Health and Safety at Work Booklet No. 22.

Only the spray gun and the connecting cables may be sited in the spray booth. All electrical ancillary equipment should be located in a separate fireproof area.

The operator's station must also be outside the booth, and the ventilation system should convey air away from the operator and towards the workpiece at a rate in excess of 0·4 metres per second.

An efficient earth connection to the workpiece is essential and the jigs and earth contact must be cleaned as a regular daily routine.

Operators should not wear gloves if they are likely to insulate the hands from the earthed handle of the spray gun. Insulative footwear must not be worn by the operator.

Powder feed hoppers should not be filled when there is a possibility of ignition occurring. Normal precautions for filling must preclude all naked lights and heated surfaces. The generator must be switched off during the filling operation.

EMPLOYERS LIABILITY TO ENSURE SAFETY AT WORK

The recent British Health and Safety at Work Act requires all employers to secure the health, safety and welfare of their employees, and to protect others from any risks associated with the products manufactured and marketed by their company. They are also required to store safely and control the use of dangerous materials and to prevent the emission of noxious vapours into the atmosphere.

Compliance with the Act can only be achieved by ensuring that all plant and equipment is safe in use, hazardous materials are defined, the necessary handling precautions are implemented and, above all, adequate training is given to all employees.

It is not yet known how this Act will be correctly applied to the British Powder Coatings industry but employers would be wise to introduce the following procedures:

(a) Define all safety hazards and inform the employees of these hazards by providing a safety manual.

(b) Employees should be asked to sign a form to the effect that they have read and understood the safety manual.

(c) Employees must be given a job instruction manual covering all the tasks they are employed to perform and training must be given to new employees and those transferring to new tasks.

(d) Routine maintenance procedures must be established to ensure the efficient operation of all safety equipment.

(e) Customers must be advised of any health risk relating to the use of the end product.

REFERENCES

1. Codes of Safe Practice—"Application of Powder Coatings by Electrostatic Spraying". Published by the Paintmakers Association of Great Britain Ltd., Prudential House, Wellesley Road, Croydon CR9 2ET.
2. "Safe Powder Coating". Published by the European Committee for Paint, Printing Ink and Artists Colours Manufacturers Associations, 49, Marie-Louise Square, 1040, Brussells, Belgium.
3. Standard SFA Part I: 1975. B.A.S.E.E.F.A., Harpur Hill, Buxton, Derbyshire.
4. British Standard BS 2506:1964, "Anti-static Rubber Footwear".
5. Selle and Zehr.
6. US Bureau of Mines Report No. 5971.

12

Testing Procedures

SUMMARY

Testing procedures are a necessity in all manufacturing operations and this is particularly true in the Powder Coatings Industry, where the high cost of the raw materials and manufacturing equipment make it essential to ensure trouble-free production and consistently to obtain a high quality, commercially acceptable, finished product.

Long before the powder coating formulation reaches the manufacturing department, the basic properties will have been thoroughly assessed in the development laboratory by testing methods designed to study fundamental flow properties during application and stoving, in addition to the performance and properties of the cured film.

Raw materials used in the manufacture of powder coatings require routine testing of all batch deliveries, though this is such a time consuming operation that it could result in approval being given (or worse still-batch rejection) after the batch has been used in production. It is a more common practice with quality control of raw materials to rely upon the extensive batch testing procedures of the resin and pigment suppliers, and to carry out occasional random checks to ensure that these materials comply with the agreed standards.

In production the plant supervisor will need a quick assessment of the colour and film flow qualities of each batch so that corrections can be made at an early stage of manufacture.

When the fine grinding operation is begun, a sample should be taken for an independent evaluation by the testing department and it may be a wise precaution to confirm that quality is maintained throughout the run by taking another sample at the end of the fine grinding process.

A sub-standard powder could contain a defect which could spoil a large number of finished articles before the defect was noticed by the applicator. Needless to say it is extremely difficult to remove the highly chemical-resistant, tough coating by conventional stripping methods.

Since it is asserted by powder coating manufacturers that their products are supplied ready-for-use and require no mixing or adjustment by the operator before application, the onus is therefore upon the powder supplier to guarantee trouble free performance, for he can hardly blame the end user for faults in a product that is claimed to be "idiot proof".

Nevertheless, the end user cannot afford to become complacent in the belief that troubles are unlikely to occur during application, and he will need to prove the quality of the cured coating, the efficiency of the spray guns, the earthing of the jigs, the quality of cleaning and pretreatment, and the stoving capability of his ovens.

In this chapter it is not possible to do more than outline the testing methods which are considered to be necessary for the characterization of powder coatings and to define the standards of performance to be expected from correctly formulated products.

TESTING PROCEDURES IN THE DEVELOPMENT LABORATORY

Much of the development work on powder coatings during the last decade has been confined to preparing the sample, spraying the powder and curing the film.

The cured film was then assessed by visual inspection and by routine physical and chemical tests to compare its performance with that of paint, an alternative coating material.

It is only recently that the fundamental properties of powder coatings have been investigated and it cannot be said that the results obtained to date have done more than point to the need for renewed activities in this field.

Work is currently being undertaken in the following areas of powder coatings characterization:

Particle Size
Particle Shape
Powder Flow
Electrostatic Properties
Reactivity of thermoset powders
Viscosity and melt flow
Pigment dispersion
Surface Profiles

Particle Size

In Chapter 8 the various methods used to assess particle size distribution were described in some detail.

These include:

Microscopic Analysis
Permeability Methods
Sedimentation Methods
Air-Jet Sieving Methods
Sensing Zone Methods

It is seen to be a relatively simple (if somewhat laborious) task to determine the particle size distribution of a powder coating sample, the problem is to determine the significance of the results.

For example, it will need to be decided how important a contribution the agglomeration of fine particles makes to poor powder flow, and can these agglomerates be easily detected during particle size analysis? Opinions are widely divided on the question of the best particle size for electrostatic spray application. Large particles have been shown to accept electrostatic charge more readily than smaller particles, and there is an increase in deposition efficiency up to a particle size of 35 microns. However, to meet the demands of the thin film coating, finer powder particles must be used from which the question arises, is it possible to apply uniformly fine particles if the coating powder has a narrow particle size range?

It has also been reported that a decrease in particle size below 75 microns does not render a powder coating more liable to explosion risk, which would seem to be inconsistent with the significant increase in surface area.

Are mixtures of fine and large particles less susceptible to air entrapment than uniform particle size powder coatings?

There are obviously many problems to be resolved in the manufacture and application of powder coatings before it can be said that particle size must be controlled within narrow limits, but it would assist powder designers if they could qualify the parameters of particle size and distribution.

Particle Shape

Mathematically, the powder particle should be spherical for ideal fluidization, packing, acceptance of electrostatic charge and uniformity of deposited film, but is the theoretical ideal the most practical shape for a powder particle?

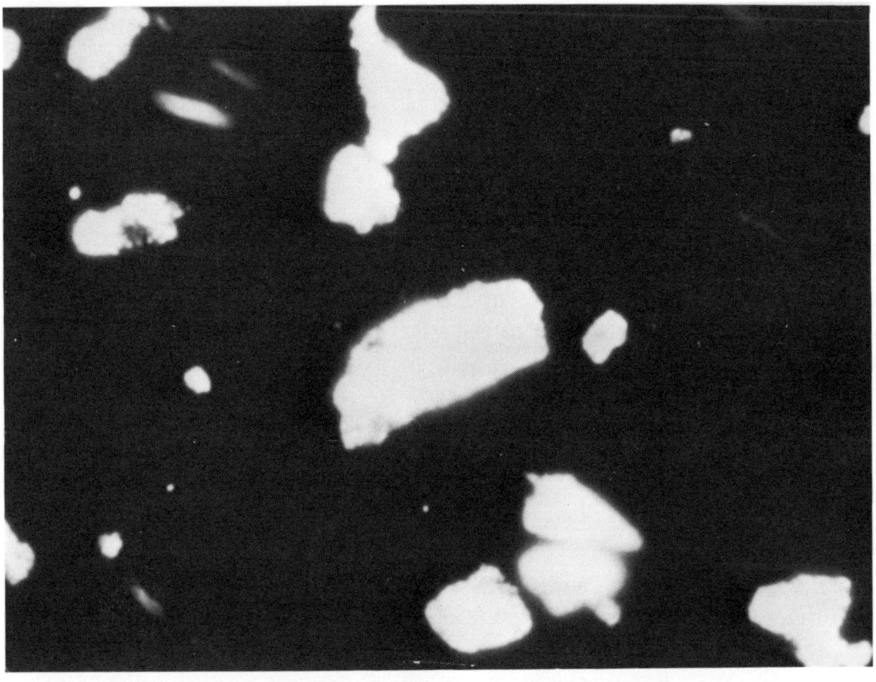

Fig. 142. Magnification of epoxy powder coating

There is little doubt that the completely irregular particle shapes obtained by current manufacturing techniques are unsatisfactory for the deposition of smooth films, and the extent of shape variation is shown in fig. 142, which is a magnification of a standard production epoxy based powder coating.

The microscope is an essential piece of equipment for examining particle shape, although techniques will have to be developed for efficient dispersion to overcome the problem of agglomeration, and the dispersion medium will have to be carefully chosen. The test when developed will be useful in quality control to check the efficiency of the fine grinding operation.

Powder Flow

Methods for comparing the flows of powders have been discussed in Chapter 8 but it is difficult to apply mathematical principles at the solid/gas interface for the gaseous phase does not assist the transmission of the forces involved in fluidity.

It would however help to define powder fluidity if a single powder was thoroughly assessed for flow, by systematically examining all the factors which are known to affect this physical property. These include particle size, particle shape, the presence of moisture, segregation of particle sizes, electrostatic charge acceptance in triboelectrification, cohesive forces, the presence of flow aids, bulking density and the readiness of powders to compact or pack down on storage.

Electrostatic Properties

It is known that powders for coating are substantially influenced during electrostatic spray application by the resistivity of the particles. This property will not only affect the ease of application, but it is also a prime factor in the adhesion of the powder layer to the substrate.

A powder with low resistivity will be difficult to apply without careful control of particle size and current density.

Powder coatings are generally formulated to have resistivities in the region of $10^{13}\ \Omega\,\mathrm{m}$.

It is easy to assess the resistivity of a wet paint for application by electrostatic spray but the same method is not applicable to powders.

The decay of charge from a deposited powder layer is, however, proportional to resistivity which has been measured by research workers at Southampton University by using a rotating field mill and observing the decay of electric field above the deposited layer.

With this principle it should be possible to develop a comparative method for the routine assessment of powder resistivity.

Reactivity of Thermoset Powders

There are very few manufacturers of powder coatings who are able to give precise information about the curing of their products with the result that the term "rapid curing" is mis-used, and widely mis-interpreted by end-users.

Klaren[1] has drawn attention to the need to investigate the rate of cure in both isothermal and non-isothermal conditions. He has examined the curing rate of powder coatings by the methods of Differential Thermal Analysis (DTA)[2] and Differential Scanning Calorimetry (DSC).[3] The DSC technique is considered to be superior to the DTA method for speed, simplicity and sensitivity and it has the advantage of providing direct quantitative measurements of heat of cure.

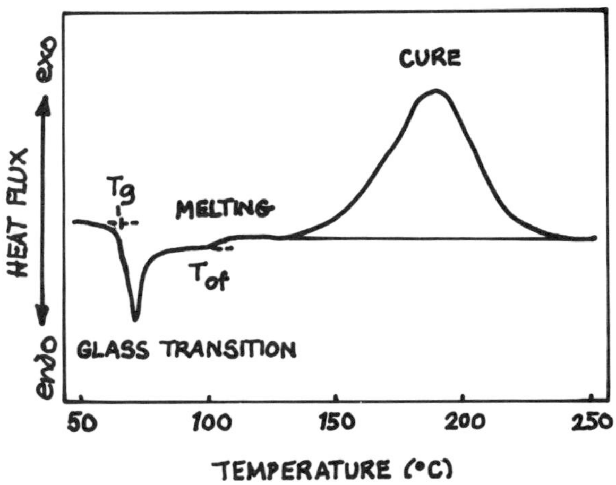

Fig. 143. DSC thermogram

These methods not only provide information about reactivity but enable glass transition points to be defined together with the temperatures of onset of powder fusion and flow.

A typical thermogram obtained by scanning experiments is shown in fig. 143 and illustrates in the vertical axis the heat flux differential between the sample and the reference, whilst heating at a constant rate. The temperature scale shown on the horizontal axis is related to the heating rate chosen.

Three effects of glass transition, melting and cure are shown on the thermogram. The glass transition temperature T_g is taken as the point of intersection of the extrapolated base line at the beginning of the endothermic shift, and the tangent to the curve at the inflection point. Cure of the powder coating is observed as an exothermic peak and the heat of reaction is measured by the area beneath this peak. Klaren reported fairly consistent heats of reaction for the samples he evaluated, in the region of 16 ± 1 cal/gm. The onset of the exotherm can be used to give comparative information concerning the reactivity of different curing agents or resins.

DSC is a useful method to employ for the examination of the effect of the heating rate upon the curing of a powder system and Klaren notes a shift of approximately 15% upwards on the temperature scale when the rate of heating is doubled.

The DSC technique can also be used to determine the stoving times required for the completion of cure, and the heating conditions during the stoving operation are simulated by combining scanning up to a peak temperature, followed by an isothermal assessment at this temperature. The results of a typical experiment carried out by Klaren are shown in fig. 144. The time taken for the DSC isotherm to level out on the base line indicates the completion of cure.

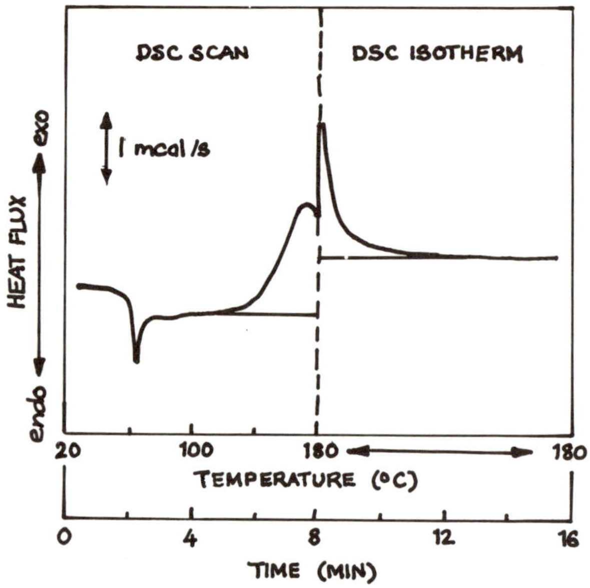

Fig. 144. Combined scanning and isothermal thermogram

Further interesting results have been obtained by Zicherman and Holsworth[4] in their examination of both thermoset and thermoplastic powder coatings employing thermomechanical analysis (TMA)[5] and DSC techniques. The TMA method records the effect of molecular weight, or inclusion of plasticizer, upon the softening points, and they reported the unusual circumstance of the lowering of glass transition temperature by increasing the pigmentation level, which also caused a reduction in the flow properties of the powder coating. They also considered the DSC experiments to be of special interest in assessing thermoset products, by observation of glass transition point and melting properties, together with the temperature range for crosslinking and the relative degree of cure.

Viscosity and Melt Flow

When powder particles melt, and flow begins, the factors which contribute to film flow are surface tension, the radius of curvature of the particles and the viscosity. Flow time has been expressed in simple mathematical terms by Nix and Dodge[6] as

$$t = \eta \frac{R_c}{6\sigma}$$

where η is the viscosity,
R_c the average radius of curvature of the particles, and
σ the surface tension.

A direct measure of the resistance to melt-flow of powder films should be obtained from viscosity readings taken during the heating cycle, and the viscosity increase of thermosetting powders during curing has been examined by use of the Weissenberg Rheogoniometer.[7]

This cone and plate instrument was chosen for its ability to heat-up rapidly to 200°C, for the small size of sample it required and for the low shear rates it applied which are important when assessing non-newtonian behaviour.

Nix and his co-workers have considered viscosity determinations with surface profiles of the films in an attempt to correlate flow with measured viscosity, and they have found examples where film flow is not entirely related to viscosity. However, effects such as this may be due to other factors, including pigment dispersion, the formation of clusters, particle size distribution and air entrapment.

Viscosity of a resin system is dependent upon the polymer structure but it may not be possible to adjust molecular geometry without affecting the physical properties of the cured film. Alterations to viscosity are best achieved by the addition of plasticizers to the resin system, but these materials will also affect the glass transition temperature.

Pigment Dispersion

It was noted in the experiments carried out by Nix and Dodge, that an improvement in pigment dispersion within a powder coating system was accompanied by a reduction in melt viscosity. This is an expected conclusion for it relates directly to experiences with paint systems, where flocculation and poor dispersion have a marked effect upon solution viscosity.

It does, however, underline the importance of cone and plate viscometry in the evaluation of extrusion procedures, of variations in performance between types of compounding machinery, and of surface pretreatments on the pigment particles. The effect of "wetting agents" in powder coatings systems can also be evaluated by these methods.

Surface Profiles

The Talysurf 4 instrument,[6] used in the comparative assessment by Nix and Dodge (see above) of surface profile in relation to melt viscosity, allows the severity of "orange-peel" to be quantitatively assessed in terms of the average vertical deviation from a centre line passing through the profile. This centre line average (CLA) is automatically computed by the instrument.

A contributory cause of orange peel is the manner in which the powder layer is deposited on the surface. Agglomerated particles can cluster together due to the presence of low molecular weight fractions, (which act as adhesives) or to residual electrostatic charge upon the individual particles. It is also noted that small particles (below 5 microns in diameter) adhere strongly to larger particles and prevent such agglomerates flowing during the stoving process.

The author has observed a significant reduction in the formation of agglomerates when particles are charged by frictional methods, a technique which is particularly effective when applying classified powder coatings with particle diameters in the range 20–40 microns.

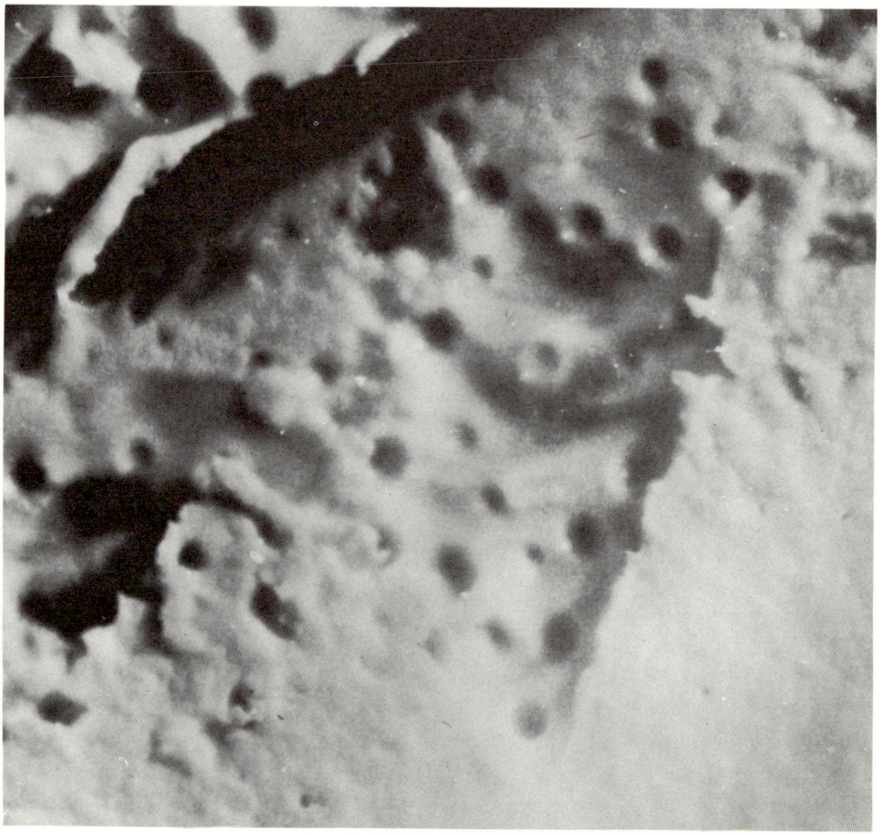

Fig. 145. The problem of air entrapment (Courtesy Scado Ltd.)

A major cause of poor film flow (resulting in pronounced orange peel) is air or vapour entrapment within the stoved film. If the coating forms a continuous film at the air interface, before air trapped in the voids between particles has been expelled, this air will be trapped just beneath the surface which will be disrupted by the honeycombed structure formed just beneath it.

This problem and its magnitude is illustrated in the enlarged cross-section of film shown in fig. 145.

There is mounting evidence that air entrapment is the biggest single contributor to poor film flow, since air trapped between the coating particles tends to raise the melt viscosity. Powder coatings applied to hot substrates give improved film flow, and additives which reduce surface tension and act as defoamers are also effective aids to flow.

It has also been observed that small particles give smoother films than large particles; this again is consistent with the reduction in voidage in small particle films and hence a reduction in the volume of air entrapped. The reported improvement in film flow with increased rate of heating is a further substantiation of the adverse effect of occluded air upon the flow of powder coating films.

Vapours driven off during the curing cycle should contribute to the problem, and polyester powder coatings ought to be more prone to suffer in this respect than epoxy powder coatings.

This in fact is the case with many polyester systems, but it is rather surprising that polyesters cross-linked with blocked di-isocyanates, (with levels of volatile by-product of up to 5%) give trouble-free, smooth coatings. It may be explained by the thermoplastic character of the film up to the unblocking temperature of 170°C and the possible contribution to low surface tension given by the blocking agent. Whatever the reason for the anomalous behaviour of the polyurethane system, it is certainly true that bubbling of the film surface is seldom observed for film thicknesses below 50 microns.

Much of the specialized equipment mentioned in this section is beyond the budget of the small powder manufacturer who will have to rely upon the Research Establishments to provide many of the answers to the fundamental problems of powder coatings. There is already a considerable effort on the part of the major raw material suppliers and it is to be hoped that successful conclusions to work currently being undertaken will result in the development of less costly test equipment to assist the powder coatings formulator.

THE TESTING OF RAW MATERIALS

It is always difficult for the small manufacturer to decide how much laboratory effort should be devoted to the non-productive routine testing of raw materials, and, if testing of raw materials is considered to be essential, what are the procedures necessary to minimize the possibility of problems arising in production.

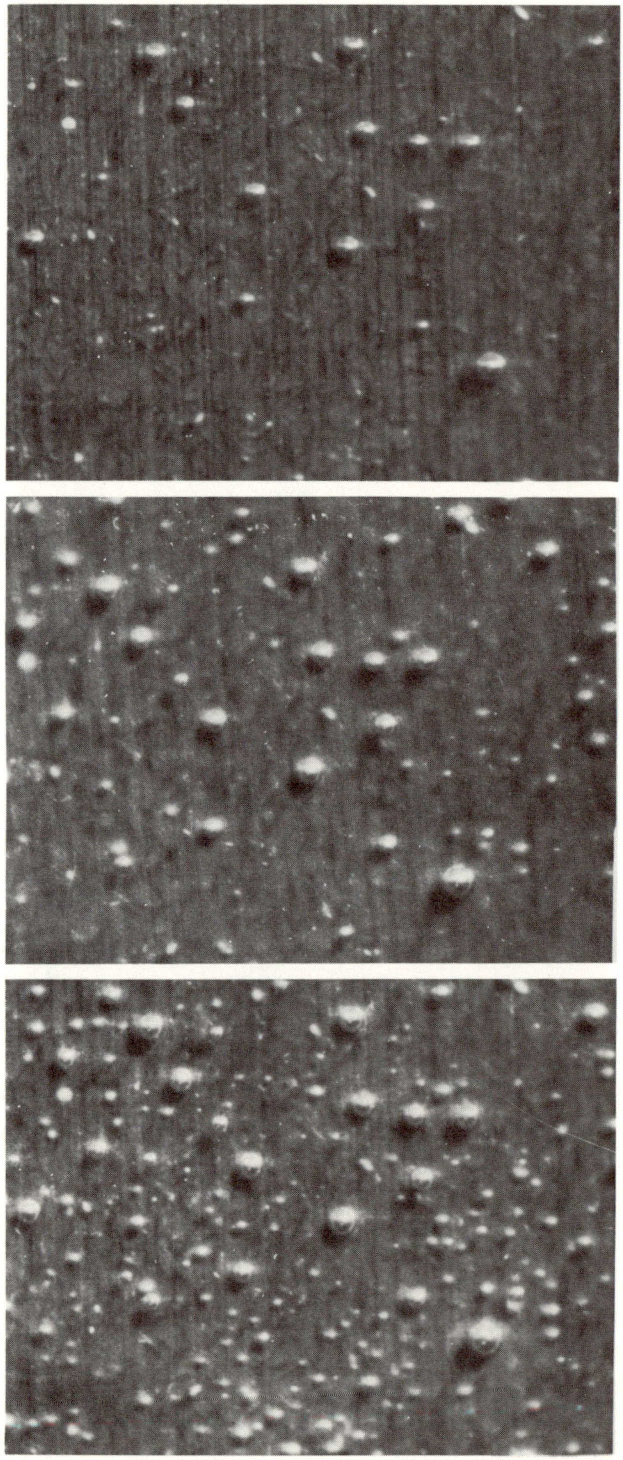

(a) 200°C for 40″

(b) 200°C for 45″

(c) 200°C for 50″

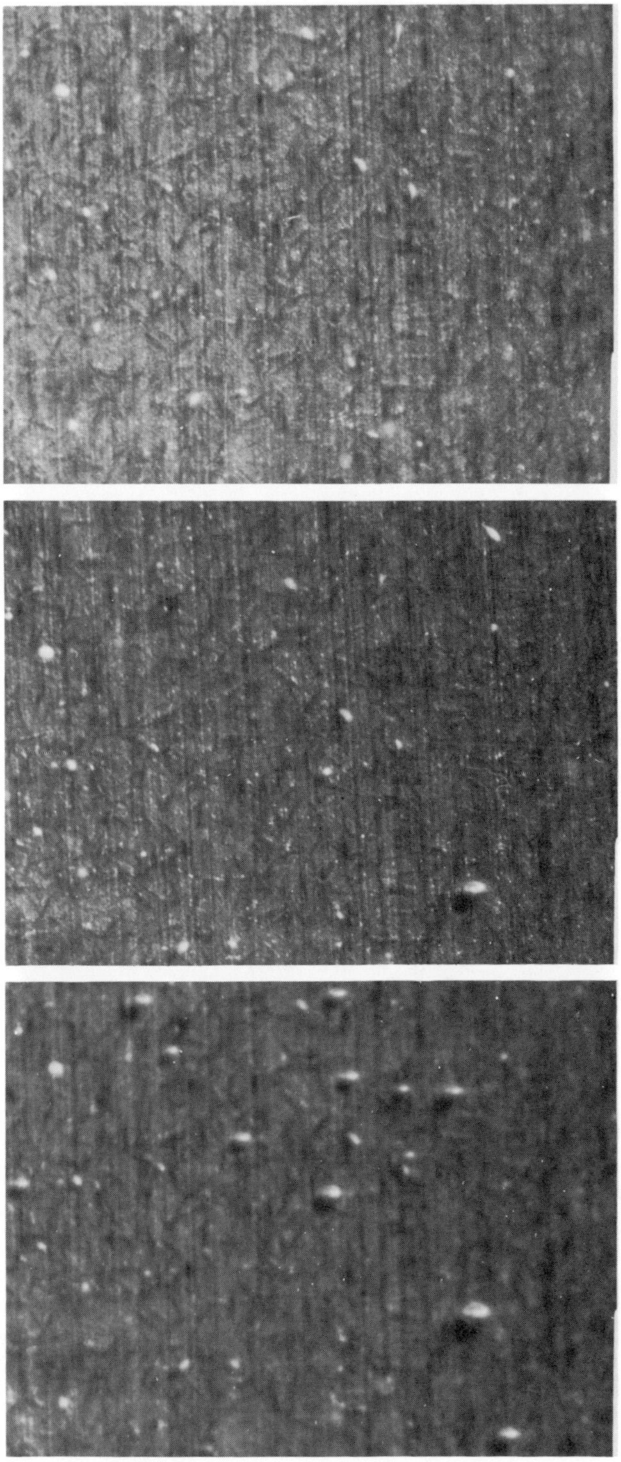

(d) 200°C for 55″ (e) 200°C for 60″ (f) 200°C for 65″

Fig. 146. Air release from the powder film during the stoving process (Courtesy Scado Ltd.)

Epoxy resins supplied by the major manufacturers' are thoroughly tested by the supplier to ensure batch to batch conformity within accepted standards.

It is not, therefore, essential to test every delivery of epoxy resin, but a routine check upon occasional batches would give a regular re-assurance to the powder manufacturer that quality was being maintained.

The simple tests required for routine checks should include:

(a) Melting range determination by the use of Durrans Mercury method, or the capillary method.

(b) Solution viscosity measurement using a 40% solution of the epoxy resin in a true solvent (e.g. the monobutyl ether of ethylene glycol) in a viscometer.

(c) A colour value reading should also be obtained with the resin solution used for measuring viscosity.

(d) Chemical assessment of the epoxide equivalent. Several methods are available for doing this and the resin supplier will generally recommend a quick and reliable procedure for solid grade epoxies. The usual procedure for the determination of epoxide equivalent is to react a known quantity of resin with a known quantity of hydrochloric acid and back-titrate the remaining acid to assess the amount used.

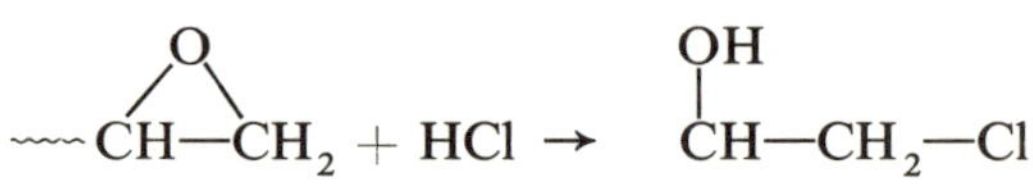

This reaction can be achieved by weighing one gram of resin into pyridine containing an excess of pyridine hydrochloride which is made by adding 16 mls. of concentrated hydrochloric acid to one litre of pyridine. After boiling for twenty minutes, the excess pyridine hydrochloride is back titrated with 0·1 normal sodium hydroxide using phenolphthalein as an indicator.

(e) The epoxide equivalent can also be measured by infra-red spectroscopy. Absorption bands at wavelengths of 10·95 and 11·60 microns are attributable to the epoxy group.

Polyester resins may also be characterized by a similar test sequence based upon melting point, viscosity measurement and colour value. Relevant chemical tests will depend largely upon the cross-linking mechanism used to cure the resin.

If the polyester contains carboxyl groups for reacting with epoxy resins or other hydroxyl-containing polymers, then the acid value will be an important test.

For polyesters used in combination with blocked di-isocyanates, it will be important to check frequently the hydroxyl value.

(The chemical methods for determining both of these are well-known and do not need to be detailed here.)

Pigments are generally of uniform composition from batch to batch and there is no real need to carry out routine batch testing.

Hardeners can vary slightly in reactivity from batch to batch but there is a considerable tolerance in the amount used to catalyse the cross-linking reaction on either side of the stoichiometric value. It is not unusual, however, for hardeners to deteriorate during storage and stocks should be kept to a minimum.

Many of the hardeners are hygroscopic and this can adversely affect anhydride curing agents by causing partial conversion to the acid with corresponding lowering of reactivity.

Flow control additives based upon polymerized acrylate esters can vary widely from batch to batch and this is evidenced by the difference in viscosity between batches of concentrated flow additive. Materials with low viscosity may be effective in the control of film flow, but they are prone to cause blooming on the surface of the coating. It is advisable to test each delivery against an acceptable viscosity standard and, when the material is approved, the concentrate should be thinned to 60% solids content by adding a volatile hydrocarbon solvent, to ensure that viscosity stability is maintained.

TESTING DURING MANUFACTURE

The important aspects of manufacture are; temperature of the extrudate, compliance with an agreed colour standard, particle size distribution of the powder and the appearance and curing speed of the applied coating.

Temperature of the extrudate is quickly measured with the aid of a thermocouple probe which should be checked for accuracy at the beginning of every production run by immersion in boiling water and recalibrating completely if necessary.

A colour standard can only be tested by actually powdering a small sample of extrudate, and applying this to a metal panel which is then stoved at the recommended schedule. The whole process should not take longer than fifteen minutes by simply grinding in a coffee mill a small amount of the granulate from the cooling band, applying the powder to a heated panel with a flock gun, and stoving in a small box oven. During this time the extruder should be operated at a low production rate. Slight variations from the colour standard can be adjusted in the premix, to which may be added the granulated extrudate which is off shade.

A simple check on the particle size from the fine grinding mill is easily obtained with a laboratory air-jet sieve, by weighing the amount retained on a 32 micron sieve.

The tests carried out prior to, and during manufacture are crude by comparison with the tests that are eventually applied to the finished product, but they are chosen for speed of operation and in the hands of an experienced operator the results are meaningful and assist him to maintain quality to the extrusion stage. After granulation, the extrudate may be stored in this form for up to twenty-four hours before fine grinding, classifying and sieving the finished product. Before the batch is subjected to fine grinding a trial quantity should be ground, sieved and tested for film flow, colour and curing properties. If the product is satisfactory then the whole batch can be ground, and a representative sample is taken from the midway stage of grinding for a more comprehensive evaluation. When a basic fault in film flow, colour or curing properties is detected during this initial assessment, it will be necessary to correct the fault by re-extrusion of the granulate with added quantities of flow aids, pigments or curing agent. It is not easy to re-extrude fine powdered materials, so it is wise to assess film properties while the bulk of the product remains in the coarse granulated form.

TESTING OF THE FINISHED PRODUCT

Every batch of finished product must be thoroughly evaluated in accordance with the agreed testing procedure before the material is despatched to the customer.

Tests included in this evaluation will largely depend upon the end-use application and the requirements of the customer's testing schedule, where specific properties may be emphasized.

In the absence of precise testing instructions laid down by the customer the following procedures may be applied to powder coatings in general.

Preparation of the Substrate

Panels (about 15 cm $\times$ 10 cm) should be closely equivalent in surface finish to the metal surface used by the customer.

Unless a pretreatment is specified, the metal surface should be degreased by wiping with paper towelling soaked in toluene. Fabrics should not be used to clean panels since they deposit fibres which are re-aligned during electrostatic spray application and spoil the appearance of the film. To ensure that panels are completely free from fibres they should be brushed with a fine nylon brush prior to powder application.

Application

The laboratory model electrostatic spray gun is easy to clean and is a most versatile piece of equipment for applying the whole range of powder coatings. Powder feed lines to the gun should be changed for the application of each type of powder coating. Separate powder feed lines should be kept, one for each of the following classes of powder coating; white epoxy, coloured epoxy, polyester, acrylic, metallic finishes and hammer finishes.

For the coating of panels, a diffuser should be fitted to the electrostatic powder spray gun to prevent the surging and blobbing of powder when the trigger is depressed.

It is found in practice that panels are coated more uniformly with positive polarity and in particular by triboelectrification devices. Equipment employing high-voltage with electronegative generators, requires careful control of voltage and powder emission to guarantee panels which are free from bits. Although the friction charging devices would seem to be the most suitable for coating test panels, it must be pointed out that all types of powder coating cannot be applied by this method, for acid cure systems will not accept charge. This would preclude some epoxies, certain polyesters including polyester/epoxy systems and several promising acrylic systems.

Stoving

Wherever possible the stoving schedule should resemble the method used by the customer and this implies that a gas oven should be used in preference to one heated electrically. Panels are best stoved by placing them directly into the hot oven set at the stoving temperature, rather than subjecting them to a gradual build up of temperature. The rapid heat up reduces the problem of air entrapment which can appreciably affect the performance of the stoved film.

Generally, the heavier the gauge of metal, the poorer will be the flow of the powder due to the extra time taken to heat up the substrate.

The stoving time quoted in any test schedule should refer to the actual time at peak temperature rather than the total time the panel is left in the oven.

Frequent checks should be made on the oven temperature by recording instruments or more simply by attaching a thermographic indicator strip to an uncoated panel placed alongside the coated panels. These adhesive strips are quite accurate, and the panel may be easily cleaned after the test.

If a powder coating exhibits poor mechanical properties it is more than likely that the panel has been understoved, for thermoset coatings are inherently brittle when undercured. Conversely, it is not easy to embrittle powder coatings by overstoving although this could affect both colour and colour retention.

When a panel has been understoved it cannot be completely cured by being subjected to a further stoving cycle. The degree of cure is inhibited at the lower stoving cycle and film rigidity of the partially cured material prevents the subsequent movement of reactive groups within the film, thereby restricting the completion of cure.

Attention to correct stoving procedures is vitally important to the performance of the coating during mechanical and chemical tests, and too often a coating is rejected in the test stations of both producer and end-user for this reason alone.

Colour

Large companies possess sophisticated equipment for comparing colour standards but this is unlikely to be available to the small manufacturer who will have to rely upon visual assessments. The eye of a practised colour matcher is an efficient instrument for comparing two standards, but it will not be able to detect slight metamerism in a daylight source.

Variability of daylight can be overcome by the use of a cabinet illuminated by artificial light. When light sources of different wavelengths are used in this device, it is possible to obtain considerable accuracy with visual colour matching.

The effectiveness of the device will depend upon the colour perception of the operator who should be examined by one of the well known methods for testing an individual's capability to assess colour.

Gloss

Gloss meters are devices which measure light reflectance from a flat surface at a set angle of incidence. Usually, this standard is 45° or 60° to the horizontal surface. The instrument is calibrated with a reference plate which must be maintained in good condition and stored in its protective case when not in use. Fingermarks on the test panel will lower the gloss readings and the operator should always handle the panels and reference plates with clean cotton gloves.

High gloss coatings are best assessed with a high angle of incidence, while matt coatings require a low angle.

The standards commonly employed in Britain are 60° angle for high gloss and semi gloss, and 20° angle for matt finishes.

Mechanical Tests

There are numerous pieces of equipment for testing the mechanical properties of a powder coating film and the testing schedule should include means for assessing the most important film properties. These are adhesion, flexibility and surface hardness.

Adhesion is most commonly evaluated by impact testing machines, employing a standard weight falling from a fixed height. The depth of penetration of the hemispherical contact attached to the fixed weight may be adjusted by altering the number of washers packed between the contact and the weight. The result is usually expressed in terms of the number of these washers which can be used to extend the contact probe, to the point where breakdown of the coating occurs.

An Erichsen impact tester records the depth of the depression in millimetres to the point where film breakdown occurs.

Reverse impact is generally more severe than direct impact.

A most severe test is to cross hatch the film by cutting through to the substrate, followed by a reverse impact at this point. Scotch tape may then be applied and the loss of adhesion recorded as the percentage of film removal when the scotch tape is torn away from the surface.

Flexibility of the coating may also be measured by impact testing but it is usual to employ a slow penetration device on a panel of lighter gauge metal.

The coating should be applied to 22 gauge metal for flexibility testing so that cracking of the substrate is not confused with film breakdown.

A bend test is a simple method for assessing flexibility on lightweight metal, by bending a strip of coated metal on a mandrel through 180°. Flexibility is expressed in terms of the radius of mandrel below which cracking of the film is observed.

Surface hardness may be measured in terms of the film resistance to a series of graphite pencils of different hardness. The pencil hardness test is, however, prone to wide variations between operators and it is seldom used except as a means of distinguishing between films of low and high surface hardness.

Reliable instrumental methods are the Konig Pendulum tester and the Tabor Abrasor.

Heat Resistance

Coated panels are maintained at a constant temperature of 120°C and rated from time to time against retained standard panels. Resistance to yellowing is quoted against an arbitrary scale ranging from 0 to 4 where

0 Excellent
1 Slight yellowing
2 Moderate yellowing
3 Severe yellowing
4 Very Severe yellowing

The test is normally concluded after 1,000 hrs.

Solvent Resistance

Coated panels are partially immersed in solvents and examined for softening of the film after one hour of immersion and rechecked for film recovery after the film has dried out for one hour.

An instrumental method for measuring surface hardness can be used to record the softening of the film.

Humidity Test

The coated panels are scored diagonally to the metal surface and enclosed in a cabinet where they are subjected to 100% relative humidity between 42°C and 48°C. An examination is carried out from time to time, and severity of film blistering is reported. At the conclusion of the test after, say, 2,000 hr the coating is stripped from half of the panel and the extent of underfilm corrosion creep is examined at the score marks.

There are many variants of this test, a particularly severe version is the exposure of the panel to humid atmospheres containing sulphur dioxide which is known as the Kesternich Test.

Table 19 Relative Performances of Commercially Available Thermosetting Powder Coatings

Resin Type		Epoxy	Epoxy	Epoxy	Epoxy
Curing agent		accelerated dicyandiamide	substituted dicyandiamide	anhydride cured	imidine
Curing cycle		190°C 10 min	180°C 10 min	160°C 15 min	140°C 15 min
Surface hardness		good	good	very good	good
Flexibility		good	good	very good	very good
Adhesion		good	good	very good	very good
Heat resistance		severe	severe	moderate	very poor
SOLVENT RESISTANCE	Aliphatic hydrocarbons	very good	very good	excellent	very good
	Aromatic hydrocarbons	very good	good	excellent	very good
	Ethyl acetate	poor	poor	moderate	poor
	Ethyl alcohol	very good	good	excellent	very good
	M.E.K.	poor	poor	poor	poor
Humidity		good	good	very good	very good
Chemical resistance of 10% NaOH		good	moderate	good	very good
Chemical resistance of 10% HCl		good	good	good	good
Accelerated weathering		poor	poor	poor	poor
Salt spray ASTM. B.11 T.		very good	good	good	very good
Boiling water		poor	poor	moderate	good
Detergent resistance		poor	poor	moderate	moderate
Volatile loss on stoving		1·0%	1·0%	1·8%	0·5%

Epoxy	Polyester	Polyester	Polyester	Acrylic
polyester	anhydride	T.G.I.C.	blocked di-isocyanate	oxazoline
200°C 10 min.	200°C 20 min.	200°C 10 min.	200°C 20 min.	200°C 10 min.
moderate	good	good	good	moderate
very good	good	very good	very good	poor
good	moderate	very good	very good	good
good	excellent	very good	excellent	good
very good	very good	very good	very good	very good
poor	moderate	poor	poor	poor
poor	poor	poor	poor	poor
very good	very good	good	moderate	poor
poor	poor	poor	poor	poor
very good	moderate	poor	good	moderate
good	moderate	poor	good	good
good	good	good	good	good
moderate	good	good	very good	good
good	good	good	moderate	poor
very good	good	very good	moderate	good
poor	poor	poor	good	poor
1·0%	0·8%	1·0%	5%	2%

Salt Spray Test

This is an accelerated test for corrosion resistance widely used in the automotive industry. Coated panels are scored in a manner similar to that used for the panels in humidity testing. Specification salt spray tests may differ in severity, due to the composition and concentration of the salt, and the temperature at which the salt spray cabinet is operated. Panels are assessed on the extent of blistering, the detachment of film from the substrate and the corrosion creep at the score mark. Most films exhibit some corrosion after 500 hr but certain epoxy powder films have withstood more than 3,000 hr without significant deterioration.

There are many more tests which can be employed to evaluate the performance of thermoset powder coatings, including numerous chemical tests and accelerated weathering. It is, therefore, important to choose the tests carefully so that they provide the right information about the performance of a powder coating under practical end-use conditions.

To simulate environmental conditions in the laboratory is difficult and a knowledge of the true performance of a coating can only be estimated by reference to one or more known standards. These standard coatings must be applied under the same conditions as the test panels, and compared at all stages of the subsequent testing programme.

A summary of the relative performance of the commercially available thermosetting powder coatings is given in Table 19.

Table 19 shows the results of a comparative assessment and relates to performance on untreated mild steel only. When the coatings were assessed on pretreated panels, the mechanical properties were poorer, but a significant improvement in chemical resistance was noticed.

REFERENCES

1. C. H. J. Klaren (Koninklije Shell Plastics Laboratory, Amsterdam, The Netherlands).
2. C. B. Murphy *et al. J. Polymer Science* **28**, 447 (1958).
3. R. A. Fava. *Polymer* **9**, 137 (1968).
4. J. B. Zicherman and R. M. Holsworth. American Chemical Society, Division of Organic Coatings and Plastics Chemistry (August, 1973).
5. J. B. Zicherman. S.M.E. Technical Paper FC 72-959. Presented at S.M.E. Powder Coating Conference, Cincinnati, Ohio (March, 1972).
6. Nix and Dodge (Sangamo Controls Ltd., North Bersted, Bognor Regis, Sussex, England).
7. Rank Precision Industries Ltd., Leicester, England.

13

Design and Layout of Powder Manufacturing Plant

SUMMARY

In previous chapters the individual stages of manufacture have been fully discussed and it will be apparent to the reader that there are many alternative choices of suitable manufacturing machinery. It is unlikely, therefore, that any two powder manufacturing units will be alike in their range of equipment or in their manufacturing capability.

The production unit must be designed to produce the output required, allow room for logical expansion, and permit the most efficient use of the equipment installed for the planned range of products.

A large prospective powder coatings manufacturer may choose to instal a plant with an initial production capacity of 500 tons per annum, based upon three or four main colours. If the number of products is considerably increased then the time spent cleaning down the high throughput machines will reduce the efficiency of the whole operation.

Preblending equipment, extruders and fine grinding machinery, all with high output rates, are costly items of capital expenditure and the manufacturer will only get a fair return on his investment if the machines are operated for large batch production.

With a target of 500 tons per annum, it is not unrealistic to aim for a minimum batch size of ten tons.

There are, however, few manufacturers who can enter the field with the assurance of a ready-made market of 500 tons per annum of powders, and it is therefore, the usual practice to aim for a lower target with smaller initial batch sizes in the hope that early achievement of the modest sales forecast will justify further expenditure to enlarge one or more aspects of production.

The volume of sales and the batch sizes achieved will largely dictate the moment to add to the production machinery.

Extruders and the fine grinding machines are continuous operation devices and their efficiency improves with an increase in the size of the production run, whereas batch repetitive operations in the preblending stage can, by comparison, become inefficient. At the preblending site small batch charges are prone to weighing errors particularly if a large number (twenty or thirty) of separate batches are needed to complete the production run. Furthermore the attention supplied by the skilled operator required could prove to be a costly labour item.

In all multi-stage manufacturing processes there must be at least one rate dictating stage. Because the extruder is the most costly item of expenditure in a powder coatings plant, it is recommended that its rate should be chosen to set the output rate for the whole plant.

The fine grinding machines, output rate should always be chosen to exceed the output of the extruder. Initially, the premixing unit should be adequate to provide in a single mixing, an hour's feed for the extruder. As total production demand approaches the maximum output rate of the extruder, the purchase of a larger volume premixing machine would be sensible for it is logical to assume that average batch sizes would also increase. To improve the versatility of the production line an additional fine grinding machine, (or even just an extra collection unit) could be purchased to complement the original equipment.

There is much to be said for a cautious approach to a production target followed by the eventual purchase of additional machinery to allow expansion within a factory unit which has been designed with space for this eventuality.

Extra machinery should be similar to the original purchased, wherever possible, to simplify maintenance and minimize the number of costly spares it is necessary to hold in stock.

This chapter merely aims to provide general guidelines to architects, engineers and production managers who may be presented with the task of installing a powder production unit.

Consideration is given to the type of building and the area, the flow line planning of production with the siting and capability of essential services, storage for raw materials and finished product, and laboratory and testing facilities.

BUILDING REQUIREMENTS

Paint manufacturers entering the powder coatings field, invariably consider that wet paint and powders should be kept well apart. This decision is, no doubt, motivated by the problems of straining wet paint to ensure freedom from bits in the applied paint film. It is debatable whether the two products need to be kept some distance away from each other because of the risk of cross contamination, and a better reason for the segregation is the fact that paint and powder manufacturing processes are quite different.

The manufacture of powder coatings is a dusty operation and it must be considered as a potential explosion risk to comply with the safety regulations laid down by the Factory Inspectorate.

Where an existing building is allocated for the powder production unit, it is unlikely to be ideally suited for this purpose, but it may be possible to incorporate some of the features recommended for the design of a new building.

It will, therefore, be assumed that a new building is to be erected to house the powder coating production, storage and development areas, with provision to expand to a production target of 500 tonnes per annum.

Site Preparation

The weight of machinery is unlikely to exceed two tonnes per square metre. An adequate water supply must be provided together with connection to foul water drainage, preferably incorporating a silt trap.

Gas should be provided although the usage is likely to be small.

An electrical power supply of 200–250 amperes should be sufficient for maximum production in the long term, with an initial demand in the region of 100 amperes.

Construction

A single storey brick walled building with no windows or ledges in the side walls which should be finished internally with smooth plaster to facilitate cleaning and reduce dust collection.

The roof should be non load bearing, lightweight, roofing sheets with translucent sections to allow maximum use of daylight. Supporting roof beams should be kept to a minimum and designed to reduce excessive collection of dust.

The floor should be smooth concrete, chemically treated or coated with polyurethane floor paint, to facilitate removal of dust by vacuum or water washing. A central drainage gully running the length of the building is an important feature. This drain should be connected to an easily accessible silt trap before discharging to the foul water drain.

Ventilation in the roof is essential for coping with the problem of heat build-up during summer months which can profoundly affect the efficiency of the whole manufacturing operation.

Dimensions

Doors must be large enough to allow easy access for forklift trucks.

The height of the building will be dictated by the height of the fine grinding collection unit which is approx. 20 ft. (6 metres).

A minimum floor area to house the production unit, storage space and development/testing facility is 10,000 sq. ft. (930 sq. metres). This total area can be sub-divided into 5,000 sq. ft. for the production department, 3,000 sq. ft. for storage of raw materials and finished products, and 2,000 sq. ft. for the development and testing department. If additional space is required eventually to increase the production area, this can be taken from the storage section by making better use of high level storage on metal racking.

Factory Heating

In the winter the heat evolved from the manufacturing process helps to maintain an acceptable working temperature for the operators, but it will need to be supplemented by heaters. Conventional radiators are natural dust collection sites and the powder will eventually fuse to the radiator and progressively reduce its heat transfer efficiency.

In the writer's experience, a gas-fired hot air blower sited at one end of the unit is the most effective method of providing additional heat regulated by a thermostatic control. Cold air should be drawn into this heating unit from outside the building to avoid collection of powder deposits within the heating device. During periods of warm weather this unit can also be employed to blow cold air to assist ventilation.

Plumbing and Drainage

Water may be required for heating and cooling extruders, Z-blade mixers, and for chilling cooling bands or rollers. It is not permitted to connect water for heating and cooling directly to the mains supply, so provision must be made to instal a water supply header tank. An alternative which offers distinct advantages is the use of a refrigerated water supply tank which can supply water for both heating and cooling. In very hot weather the cooling efficiency is considerably reduced by the increase in ambient temperatures of the water supply, and the output rate of the extruder may be drastically lowered. To ensure a fixed cooling water temperature of 6–8°C during summer and winter, a centrally situated refrigerated tank can supply several machines and the extra capital outlay is recompensed by maintaining extruder efficiency.

A mains supply of water is also required for cleaning down grinding machinery and collection units and for floor cleaning. Water taps sited at frequent intervals with hose connections will greatly assist the cleaning down of equipment between batches. It is not suggested that large amounts of dust will be washed into the drainage system, for the water wash is only applied after the machine has been cleaned down with a good quality industrial vacuum cleaner. Traces of powder left in inaccessible corners are then removed by hosing down the machinery. In a similar manner the floor would first be cleaned by vacuum followed by water washing to complete the cleaning operation. The relatively small amounts of powder washed into the drainage system are unlikely to cause any effluent problem, but the provision of a silt trap is an additional safeguard. Water for cleaning purposes would not be taken from the refrigerated unit.

Electrical Fittings

It cannot be stressed too strongly that dust collection is the basic problem in powder coatings manufacture and electrical switches and connections must be made as dustproof as possible. Flame proof fittings are to be avoided for their mode of operation would allow powder to reach the contacts.

The main fuse board should be totally enclosed and preferably sited in the storage or development areas where the dust collection problems are less than those in the production unit.

Trunking carrying the ring main power supply should have the facing panel secured with the maximum number of holding screws and junction boxes for conduit sections must have well secured cover plates.

Push-button starting switches are notoriously susceptible to the ingress of powder coatings which "weld" the contacts together. It is, of course, possible to purchase dustproof push-button switches and these should be used from the outset.

Most industrial vacuum cleaners are fitted with push-button switches and these will fail within a short space of time. If these are replaced by the conventional lever switch, there is less likelihood of failure.

Purchased machinery is invariably supplied with its own electrical control system and these are generally trouble-free in operation.

Three-phase junction boxes do not give problems but it is advisable to provide the connection socket with a screw fitting cap to prevent dust entering the contacts when the socket is not in use.

When installing production machinery care must be taken to ensure that the machine is safely earthed. Cooling bands, granulators, pre-blending machines, fine grinding machines and dust collection units will generate electrostatic charge during operation, due to the triboelectrification effect and this charge must be conveyed to earth.

Fluorescent strip lighting should be encased and cleaned regularly every month. This is the most suitable form of artificial lighting and does not seem to be greatly affected by the dusty atmosphere of the powder manufacturing unit.

FLOW LINE PLANNING

Although the initial installation will not be operated on a flow line basis, it would be prudent to plan for the possibility of continuous production and ensure, at the outset, that the essential machinery is correctly sited to facilitate a trouble-free conversion at a later date.

The sequence of operations during the manufacture of powder coatings begins at the weighing station and is followed by preblending, extrusion, cooling and granulating, fine grinding and classification, sieving and finally, packing and weighing the finished product.

A gravity flow line would be the ideal set-up for continuous production of powder coatings but this would need a three storey building, and the cost of such a building would be a high capital committment. Simple devices for transporting and elevating the product between production stages are relatively inexpensive and allow flow lines to operate at ground floor level.

The Weighing Station

This will be sited in a position convenient to the main raw material storage area and adjacent to the preblending machinery.

A single set of scales, capable of weighing one hundred kilograms with a tare of twenty kilograms, should be sufficient for the initial installation since the bulk of the preblending charge will be measured in the weighed packages used to supply the raw materials.

While the initial batch size will be relatively small, an increasing demand will be placed upon this section as batch sizes increase. It will therefore require some organization to ensure that individual weighings are accurate and uniform throughout the production run.

At the end of each production run the weighing area must be cleaned and cleared of empty packages and remnants of raw materials used during the sequence of weighing operations. If the next production run requires a large number of individual preblending batches, then the flow line production technique will be assisted by prior planning at the weighing station to establish a routine procedure and facilitate the individual weighing process.

(i) The raw material store must be issued with a request to supply the *total* quantity of materials required in the production run, well in advance of the needs of the production department. This total should allow an excess of between 3 and 5% above the desired yield to cover losses during manufacture. The excess amount will depend largely upon the batch size, but even with the small batch it is unlikely to exceed 5%.

(ii) Before beginning the first weighing operation, the total weight of raw materials supplied by the stores should be checked by the operator against a copy of the stores requisition formulation.

(iii) The size of the individual charge will be dictated by the size of the preblending machine but the total charge must be a simple multiple of the individual charge to avoid the troublesome operation of weighing a part batch to complete the charge.

(iv) The largest quantity of raw material used in each charge will invariably be the base resin component and it will help the weighing operation if the individual charge is based upon a whole number of suppliers packages thereby avoiding the need to weigh out the resin component.

(v) A procedure which will considerably help the operator to weigh accurately the small amounts of pigments and flow additives, is to pre-mix the total quantity of all these ingredients so that he can then take an aliquot of the mixture for each weighing. A simple drum mixing device is suitable for the premixing of these small amounts of raw material, and it will be found that good batch to batch consistency with coloured coatings can be obtained and the repetitive tedium of repeatedly weighing small quantities is avoided.

(vi) If the total amount of ingredients has been weighed out in the stores before individual batch weighing commences, stock records can be debited by the stores department and it will be unnecessary to produce individual batch sheets for each weighing or even to signify on the batch sheet the number of weighings of individual batches.

A check on the accuracy of weighing will be evident after the last operation when all ingredients should have been used up.

Preblending Machines

These have been previously described in Chapter 5 and the reader will be aware that there are many alternatives that can be chosen.

To plan for a "flow line" production facility in the preblending area it is necessary to consider carefully how the next stage of the manufacturing process is likely to develop with increasing output.

If preblending is carried out entirely in 200 litre drum hoop mixers, then it will not be important to consider a developing demand in the extrusion section, for this simple blending equipment is portable and its output is easily increased by providing enough drums of premix to satisfy the rate of extrusion.

Since each drum is an individual container for premix, it would be possible to cater for the batch requirement of the extruder by preparing the premix well in advance and storing drums of premix in a "holding station" until the production run is begun. This area could be provided in the stores section to avoid taking up too much space in the production zone.

With permanently sited preblending machines it will be found that a batch of premix can be made in about five or ten minutes from the completion of the charge. Therefore, a relatively small volume capacity preblending machinery can easily keep pace with the continuous requirements of a single extruder. The problem with one preblending machine will arise when several extruders are to be supplied from the single machine and producing possibly more than one type of premix. Even in the face of such a forbidding demand it would be possible for the single preblending machine to cope, if necessary, by providing a "holding station".

If the maximum efficiency is to be gained from the use of the preblending machine, it must be used for mixing only with the contents being quickly discharged so that the equipment is never used as a storage vessel for longer than is necessary to process the premix.

Containers suitable for holding and storing the premix should be designed so that they will fit beneath the discharge point of the preblending machine and be capable of connecting directly to the discharge outlet to minimize the dust evolved in the emptying procedure. They must also be constructed for easy handling by forklift truck and be capable of stacking one on top of another to save storage space.

A bottom emptying outlet should be provided in the holding tank if it is required to discharge directly into the feed hopper of the extruder. There are several ways by which the premix could be transferred from these containers to the extruder feed hopper including; direct coupling of the tank to a screw feed elevator, pneumatic transportation in a closed system or, by simply raising the tank by forklift truck, above the feed hopper and discharging directly by gravity.

It is important for "flow line" production that separate premix charges should be blended in a consistent manner throughout the entire production batch. To ensure uniformity of premix a preset timing switch should be included in the controls. This will be even more important if a high-speed mixing device is used to preblend components, for heat is generated rapidly by the high work load used in the mixing operation. High-speed machines are generally provided with two stage mixing compartments so that the heated charge can be transferred to a cooling section before it is finally discharged. This two stage mixing machine is usually very costly to operate and it would be uneconomical to employ the cooled compartment as a "holding station" for premix.

Correct siting of the initial preblending machine is important to permit this area of production to be expanded as the demand increases. The approximate dimensions of the total production zone will be 100 ft. (30 metres) long by 50 ft. wide (15 metres) with the flow line proceeding logically down the length of the workshop. The first preblending machine should be sited almost midway across the shop, adjacent to the central drainage system and close to the weighing station. This will allow the machine to be accessible to more than one extruder and permit expansion of the preblending site on either side of the initial installation. It will also provide the maximum amount of working space round the machine which will be an important consideration during the early days of production when the problems of handling and conveying the premix have still to be resolved by a certain amount of trial and error.

The preblending operation is possibly the most important of all the production processes because the quality of the premix is reflected in the performance of the finished product. A coarse, heterogeneous blending of the components cannot be ideally homogenized in the short barrel length of an extruder during the average dwell time of thirty seconds. If the extrusion

Fig. 147. Typical preblend

process is to be expected to provide some of the mixing which should have been carried out at the preblending stage, then the rate of extrusion will be adversely affected and the production efficiency will suffer.

The structure of typical premix is shown in fig. 147 and the problems of blending epoxy resin "as supplied", with finely ground pigments are made apparent.

To obtain the best combination of good dispersion and good throughput rate at the extrusion stage, it is recommended that epoxy resins are granulated to 2–4 mm diameter particles before preblending.

The Extrusion Process

This should be the rate dictating process for flow line production and the machine must always be operated at control settings which afford the highest throughput of premix consistent with efficient dispersion. It will not be possible to process all premix formulations at the same control settings and it will be necessary to carry out several trial runs with new products to arrive at the most suitable machine settings.

The factors which will affect the efficiency of extrusion are:

(i) the screw speed
(ii) the feed rate of the premix
(iii) the granular size of the premix
(iv) the composition of the premix
(v) the design of the feed section
(vi) the settings for the heated zones
(vii) the design of the extruder outlet

(i) At the beginning of the extrusion process the screw will be operated at a relatively low speed until the barrel is completely filled with premix and hot melted extrudate is being continuously discharged. When this moment is reached the screw speed is then increased to its optimum value which is invariably the maximum speed setting.

Screw speed is the one control setting which should not be varied with each new product.

It will be found that the maximum screw speed will be considerably different in each type of extruder. A planetary extruder will operate at a lower screw speed than will a single screw machine which in turn has a lower maximum speed than twin screw extruders.

(ii) The feed rate is generally controlled by the speed of the paddle in the feed hopper or, in the case of vibratory feed systems, by increasing or decreasing the amount of vibration in the feed trough and hopper.

As the rate of feed is increased the work load on the screw will also increase and the optimum working conditions can be controlled by reference to dials showing either the power consumed or, better still, the torque. It will greatly assist the operator if the feed control is directly related to the reading on a torque meter to avoid the fluctuations which could occur with occasional surging of the feed and to allow the introduction of a third control—the back pressure generated by adjustments at the discharge point.

The value of operating the feed directly from the torque meter will become apparent during a long run because the stabilization of temperature conditions within the machine after running for some time would allow the feed rate automatically to increase as the work load was slightly reduced. Furthermore, the slow build up of partially cured product in the die head would restrict the outlet and eventually overload the machine if the feed rate was not reduced. This simple device could enable one operator to supervise several machines and it is perhaps the most important control feature in the "flow line" operation of the extruder. It has previously been recommended that the feed to the extruder should be connected to the torque meter to avoid the possibility of damaging the gears in the event of an overload on the extruder.

(iii) The particle size of the granulated premix will affect the throughput rate, since the extruder must be capable of conveying premix from the feed section into the main body of the machine where hot melt compounding can begin. If additional work is still necessary in the feed section to break down large lumps of resin, then the rate of feeding premix will have to be decreased to prevent overloading and it may be necessary to increase cooling efficiency in the feed section to avoid premelting brought about by mechanical heat evolved in the granulation of oversize resin particles. This additional work load could also cause melting of the resins to occur closer to the feed section which could get progressively nearer to the inlet with consequential loss of extruder efficiency and even a complete blockage of the feed compartment.

A finely powdered premix will also affect the rate of extrusion adversely because the premix will then have a lower bulk density; but an effect perhaps even more important than the increased volume of the premix will be the ability of the feed screw to grip the finely powdered material and convey it against the back pressure in the hot melt section. The particles in these circumstances are of course much smaller than the clearance between the screw and the barrel. It is found in practice that it is difficult in single screw extruders to convey powdered premixes due to the high back pressure in the barrel and it is not unknown for powdered premix to be forced back from the feed section into the gland where it can appreciably reduce the extruder efficiency because of the thermoset characteristics of these products.

Twin screw extruders can handle powdered premixings but at a reduced rate of extrusion.

Since the major component of the premix is the base resin, it will be found that the best feed rates are shown when the resin particles are granulated to between 2–4 mm diameter prior to the premixing operation.

(iv) The rate of extrusion will also be affected by the composition of the premix, the rheological properties of the ingredients and the relative quantities used in formulation.

Large amounts of extenders and pigments, particularly titanium dioxide, can effectively reduce the throughput rate for several reasons. These materials have a very small particle size, in the region of 2–5 microns for extenders and sub-micron size for titanium dioxide.

Small particle-size can adversely affect the rate of extrusion for the reasons given in the previous section; fine particles also agglomerate readily and cause blocking at the inlet to the feed compartment. Pigments with high oil absorption values are not easy to disperse and it will be necessary to increase the dwell time in the compounding section when high concentrations of pigment and extender are present.

Channel blacks are usually employed in powder coatings at low concentrations but even at such levels they are extremely difficult to disperse. When they have to be used, either the rate of extrusion must be reduced to allow additional time in the compounding section or the back pressure must be increased to raise the work load which in turn reduces the feed rate. Often the inclusion of some extender can improve the dispersion of carbon blacks but a better alternative is to predisperse this pigment by the conventional method of chip manufacture.

In fact, all strong coloured pigments are best incorporated at the extrusion stage in a predispersed form by a masterbatch technique.

Resins employed in formulation can vary considerably in their melting points, but the more important qualities of the resins are the melt viscosity and the ability of the resins to wet the pigment particles.

Some work on the assessment of melt viscosity and wetting properties in relation to the applied film has been carried out by Gabriel (1) but this work has not been translated to the manufacturing process.

Wetting agents are useful components of powder premixings for they can markedly assist pigment dispersion.

(v) Several manufacturers of well known extruders have re-designed their feed screws to improve flow rates of powder coating premixes into compounding sections. It has been appreciated by the extruder manufacturers that the problems of powder coating extrusion differ from those of plastics extrusion. Improvement in feed rates has been mainly achieved by altering the angle of the leading edge of feed screws and by reducing slightly clearances between the barrel and screw in the feed compartment. However, the greater thrust of the modified feed section does tend to pack the feed more tightly along the leading edge and this can drastically reduce the rate of feed after the machine has been running for some time.

Feed sections of most extruders are water cooled and the precise control of temperature in the feed compartment can make a significant contribution to the efficiency of feed.

Obviously, a high temperature in this zone will promote agglomeration of resin particles and cooling systems must therefore prevent the build-up of excessive mechanical heat. A surprising feature is the falling off in rate of extrusion observed with some single screw machines when the cooling water temperature fails below 6°C. It is not understood why this reduction in feed rate should occur unless it results from a build-up of back pressure in the barrel brought about by the melting of the cooled premix at a point which is now further away from the feed section.

In the event of blockage in the feed section markedly affecting the rate of extrusion, it will be necessary to clean out the feed compartment. To minimize delay caused by this kind of cleaning operation, the feed section must be designed to allow easy access for cleaning.

When the extruder has been in operation for some time it will be noticed that wear will begin to appear in the screw and the barrel, particularly in the feed section and the adjacent zone which is before the premix has melted. As this wear progresses the gap between the barrel and screw will increase and the extrusion rate will suffer.

(vi) The manner of heating varies with type of extruder but the zones needing heating will be approximately the same in all machines. A separate heating unit and circulating pump is provided for each of the heating zones and a thermostatic control enables the temperature to be set within precise limits.

Earlier models of commonly used extruders were provided with oil circulating heating systems, a method which is no longer favoured because of the indirect cooling it needs and the relatively poor heat transfer efficiency of the medium. Electrical heating is much less cumbersome than a circulatory system but cooling is difficult and it has been found in practice that such fierce heating has an adverse effect upon the quality of highly reactive powder coating formulations.

The preferred method of heating is a water circulatory system which provides good heat transfer and highly efficient cooling. Powder coating manufacture does not require high temperature heating and hot water is now sufficient for most extruder operations.

Heating may be applied to the screw and it does assist the stabilization of temperature during extrusion if the screw is warmed before beginning a run. A temperature in the region of 50–60°C should be adequate for most processes. If a fault should develop with the screw heating system during a production run, the extruder will continue to function with a minimal reduction in efficiency if the water circulating pump is switched off as soon as the heater fault becomes evident.

The several zones of the barrel are usually heated separately and a temperature gradient will be maintained as the material progresses down the length of the barrel. It is important to take the advice of the extruder manufacturer in the initial stages of operation since his recommended temperature settings will ensure that the product melts in the right zone of the barrel. Their advice will be based upon figures previously decided for fairly standard powder coatings. As the operator gains experience with the machine it will be possible to make slight adjustments to the barrel heater settings to improve compounding efficiency or to increase the throughput rate. Barrel temperature settings are usually about 10°C below the melting point of the resin.

Correct temperature control is essential at the die-head or discharge outlet of the extruder. Manufacturers are now aware of the superiority of circulating hot or cold water for heating or cooling the screw and barrel yet some of them still persist in supplying electric band heaters for the die head where correct heating or cooling are perhaps even more important. The additional heat provided by the band heater promotes gelation of the reactive extrudate and the significant build-up in the die then results in lower throughput rates. Water heating is the only effective method for controlling temperature at the die head and discharge zone!

(vii) Many powder coatings manufacturers resolve the problem of material curing in the die-head by the simple expedient of leaving the die off the machine during the production run. This technique is adequate for the production of standard medium quality powder coatings but it is not recommended for the extrusion of high quality products. Removing the die is a simple method of increasing the outlet but it also affects the back pressure which is important when a high degree of dispersion is required. When a planetary extruder is operated the die must be in place to prevent the planets being ejected from the barrel. An increase in back pressure can only be obtained by replacing the standard die with another die possessing a smaller outlet aperture.

This comment is generally true for all single screw extruders although the majority will operate without the die in place. It is important for ease of cleaning between batches to have several dies available so that the coated die can be cleaned by immersion in a stripping solution rather than being subjected to mechanical cleaning which can pit and damage the die. If dies could be easily modified by the addition of different sized annular-ring outlets, this would enable a better degree of dispersion to be obtained with formulations requiring efficient compounding. With increasing interest in thin film application of powder coatings, the refinements suggested in this paragraph could offer an inexpensive way to improve the efficiency of early extruder models.

Certain models of twin screw extruders have variable discharge orifices which avoid the necessity to restrict the die exit when an increase in back pressure is required. Variable discharge orifices can be progressively adjusted between "open" and "closed" extremes by mechanical adjustments of the screws in relation to a venturi shaped discharge block. These alterations may be made at any time during the production run without interrupting the process.

It was suggested earlier that preblending machines should be centrally situated in the production department so that each can serve more than one extruder. To permit this, the initial extruder must be sited approximately one quarter of the distance across the production area with the direction of flow away from the weighing areas and preblending site. The cooling band,

and the granulation device at the end of the cooling band, are already suitable for flow line production and it is only necessary to instal a method for either conveying the granules to the fine grinding machine or for, storing the granules in a "holding station" in a similar manner to that described for storing the premix.

The Fine Grinding Machinery

As described in Chapter 7 this will require careful siting if it is to complement the basic aim for flow line production. Small throughput machines with a capability of 50–150 kilos per hour are best sited along the walls of the production department so that the central area is left free for expansion in high capacity machines. These small machines are necessary for they will easily cope with trial quantities and small orders for specific colours, without the time consuming cleaning procedures of the large collection units.

A large throughput fine grinding machine should be part of the initial capital outlay on powder producing equipment and it is this grinding machine which should be installed with expansion in mind.

If, therefore, a pin disc mill with variable speed classifier has been purchased it will probably be used in conjunction with a dust collection unit and at a later date be operated with an in-line classifier between the mill and the collection unit.

The mill will have a greater production rate than the extruder for the whole production set up is related to the pace of the extruder. Cleaning the mill between different coloured batches is relatively simple and should take the operator little more than one hour to accomplish. The collection unit on the other hand is difficult to clean down, especially if the change is from white to black where small amounts of residual powder could easily contaminate a large quantity of powder coating. Expansion in production will require the manufacture of large runs of white and possibly black or grey powder coatings. Ideally a separate mill and collection unit for each colour would greatly facilitate "flow line" fine grinding of large batches but such items are quite costly and there is bound to be an interim period during the growth of production when the complete cost of a new installation could not be justified. With this frustrating period of expansion in mind it would be wise at the outset to maximize the work capacity of the fine grinding mill. This can be achieved in some measure by installing a portable pin disc mill. In practical terms it would be difficult to lift and resite a heavy mill but a machine mounted upon casters can easily be moved along heavy gauge metal channel to an adjacent site. At the second site provision would be made for the eventual installation of a further collection unit as it would be a simple matter to connect the mill to the new collection unit with flexible trunking. The metal channelling should be laid parallel to the direction of production flow with the collection units sited along the wall adjacent to the smaller and equally permanent low capacity grinding machines. The section of roofing above these collection units and grinding machines could be elevated if necessary, with explosion relief ducting established in one relatively small area of the roof. This would assist maintenance, and any further precautions required by the Factory Inspector would be less costly to install.

The in-line classifier could also be mounted on casters and be moved in conjunction with the grinding mill.

A less satisfactory alternative to this scheme is the permanent siting of the mill in a central position of the production area so that collection units can be grouped round the mill. This has the advantage that the laborious task of moving the mill can be avoided, but the grinding area becomes very congested and it would be a serious problem to convey granulated feed to the grinding machine. The original scheme allows the holding tanks of granulate to be moved freely down the "line".

The Final Area in The Production Site

It is where powder sieving can be undertaken although this operation is usually considered to be an integral part of the fine grinding process. Many manufacturers do not bother to sieve their finished product even though there are advantages to be gained when the powder is discharged directly from the collection unit. For example small amounts of fibre from raw material packages and lint from collection filter bags can be removed, and warm agglomerates are cooled and broken down by the sieving operation.

Powder discharged from sieves or directly from collection units or in-line classifiers will be packed into boxes containing polythene liners, which will then be conveyed into the storage and despatch area. It will assist when planning the storage area to know that one metric tonne of powder packed in 25-kilo cartons will occupy a volume of 2·3 cubic yards or 1·758 cubic metres.

LABORATORY AND TESTING DEPARTMENT

It has been a general principle in the Paint Industry that the person who formulates a new product should not be the one to test and approve that product. This procedure of independent arbitration is also advisable in the powder coatings department.

In the manufacturing establishment being considered, an area totalling 2,000 sq.ft. was allotted for development and for the installation of testing facilities and a subdivision into two areas of equal size should allow adequate space to house the equipment and activities of each section.

It is not proposed to detail the items of routine furniture and shelving but to concentrate on the essential pieces of equipment together with the ancilliary devices which will enable the equipment to be used efficiently.

The equipment needed in the development laboratory will be similar to but smaller than that in the production department. Batch sizes will seldom exceed 2 kilograms but the relatively large number of development formulations to be processed will involve the technician in a considerable amount of cleaning between each development run.

Small versions of the production preblending machines are available, but their volumes are seldom smaller than one cubic foot which is too big for handling two kilogram samples; the machines are also difficult to clean. The most suitable laboratory sized preblenders are food mixing devices which can be bought from most domestic appliance manufacturers. These readily available mixers are cheaper than the specially designed laboratory machines and spare parts are not difficult to obtain at a reasonable cost. It cannot be claimed that these low cost mixing appliances are ideally suited for preblending of powder coatings for they are not sufficiently robust in construction. There is a need for the development of a special mixing device using the principle of the food mixer but with a more powerful electric motor, than is usually fitted, an efficient dust proof gland, robust metallic paddles and a metallic container and lid.

An alternative method of preblending (and one which is quite adequate for mixing small samples) is the simple process of tumbling all the ingredients together within a closed container. A five litre open-aperture can will easily contain the two kilogram sample charge and, after sealing, the ingredients are mixed by rolling the can for about five minutes.

A small capacity laboratory extruder, capable of handling between 10 and 15 kilos per hour, is essential for development work. The Buss Ko-Kneader PR 46 is recommended because its output rate is high enough for sample manufacture yet not high enough for it to be commandeered by the production department as a small batch production unit. This machine should be water heated and cooled, and possess a split barrel to facilitate the cleaning down operation. It is not necessary to attach a cooling roll to this machine, since a simple cooling table is quite adequate.

A laboratory size granulator will be required, to reduce the cooled extrudate to a size suitable for feeding to the fine grinding machines, and this piece of equipment should be chosen with ease of cleaning as the prime design feature.

The choice of a laboratory fine grinding machine is fairly restricted for many manufacturers do not include a small machine in their range of grinding equipment. An easy machine to operate is the Mikropul Bantam which is a hammer mill capable of grinding twenty kilos per hour. It is possible to exercise some control on the particle size by choosing a suitable sized screen through which the particles leave the mill. The Bantam is not difficult to take apart, clean and reassemble, but it is noisy in operation and should be installed in a sound proofed compartment.

If it is required to grind a few grams of material quickly for the purpose of checking a production shade or quality, this is best achieved in a coffee grinder supplied in a food mixer kit. Less than one minute is enough to prepare a small sample for application and testing.

Laboratory fine grinding machines do not produce the fineness of grind obtainable in production and the flow of the applied sample will generally be poorer than the standard which could be obtained in large scale manufacture.

A small sieve will help to improve the application characteristics of the sample but it is not easy to sieve powders with particle sizes much finer than eighty microns.

Since a fair amount of laboratory effort will be devoted to cleaning equipment down between samples, it will be advisable to instal a solvent-cleaning bosh where machine parts can be immersed.

The remaining main items required in the testing area are the application equipment, the spray booth and the ovens.

In Chapter 10 there is a full description of most makes of electrostatic powder coating handguns and the purchaser is advised to take advantage of the test house facilities provided by these manufacturers before ordering spray equipment and booths. To cater for the demands of the sample development section, the routine testing of production batches, preparation of customers specimens in accordance with sales requests, and technical service visits to apply powder at the potential customer's premises, it will be necessary to purchase more than one electrostatic hand gun.

There is some advantage to be gained by purchasing both guns from the same supplier for most standard generators have attachments for two guns which could save up to 30% of the cost of two separate installations. However, it is generally accepted that two different makes of hand gun can better serve the needs of the testing department for there are considerable differences in performance between the various types available.

If it is decided to produce specimens in the testing department then a large open spray booth should be purchased together with an oven capable of stoving the majority of customers specimens.

The manufacturer of the spray gun is the best person to consult about the design of the spray booth and he will advise the manner and method of construction if a custom built booth is required.

A large gas fired box oven with air circulation is the most suitable equipment for stoving powder coated specimens.

A list of manufacturers of booths and ovens is included in the appendix.

Specific items of testing equipment have already been described in Chapter 13 and their inclusion in the testing section will depend upon their usefulness for assessing compliance with specifications.

REFERENCES

1. "The flow of epoxy based powder coating films in relation to reactivity, rheology and wetting", by S. Gabriel (Koninklyke/Shell Plastics Laboratory, Delft, Holland). Presented at the Oil and Colour Chemists Conference in Scarborough during June, 1975.

14

Design and Layout of Powder-Coating Installations

SUMMARY

To help the potential end-user of powder coatings to choose and lay out equipment, this section describes the items of application plant needed for hand operated units, for semi and for fully automated units. A comparison is made where possible between systems for applying powder coatings with the equivalent wet paint systems.

Attention is drawn to aspects of powder coating which have been discussed in more detail in other Chapters, yet need to be recounted here so that their place in a powder application plant can be fully appreciated.

It is not suggested that these remarks should be more than general guidelines to be discussed in greater detail with equipment suppliers, who can determine the exact requirements of the customer from trials carried out with the actual articles to be coated and from a knowledge of the volume of production required.

Examples of existing end-user plants are included in the text together with photographs showing installations and coating techniques.

THE ARTICLE TO BE COATED

When first considering the application of paint coatings by spraying powdered paint electrostatically, it must be quickly determined if it is feasible to apply coatings effectively in this way to the manufactured article or component. In fact most manufactured articles present little difficulty and a trial carried out in the test house of an equipment supplier usually provides a pleasant introduction for the prospective end user, to the simplicity of the coating process and the high quality of the finish obtained.

The article to be coated need not be metallic, for there are already plants operating for coating electrostatically black phenolic plastics with powder, where a single coat of powder gives a film superior in performance to that from a two coat wet-on-wet polyurethane system applied conventionally. At the same time it provides a significant reduction in reject rates.

Metallic articles are easily coated in one spray operation with a film thickness in the range two to three thousandths of an inch (fifty to seventy-five microns) and, while this is an acceptable alternative to a multi-coat wet-paint system, it is not an economic proposition for the manufacturer who is currently applying one thousandth of an inch (25 μ) dry film thickness by a wet paint, one coat, process. However, even this valid objection is unlikely to be substained now that powder coatings are commercially available to provide thin films.

Box section components have in the past been difficult to coat successfully with electrostatically sprayed powder coating due to the electrostatic repulsion of the so-called "Faraday Cage" effect, but here again technological advances in the design of spray equipment have made it possible to minimize this effect. Powders are now capable of coating the difficult recesses more efficiently than wet paint processes in which the film in the recesses suffers from solvent washing.

Hammer finishes with a uniform pattern can be produced by spraying silicone free powder coatings. Large articles can be coated without encountering the problem of film-sagging on vertical surfaces, a problem often experienced with wet paint hammer finishes. The same comment is true of textured finishes and other types of non-conventional coating where the success of a wet paint application is entirely dependent upon paint viscosity, the skill of the paint sprayer and the flash off period.

With powder coatings there is no flash-off period, no running away from sharp edges, nor sagging due to excessive film thickness or secondary flow.

The ready-for-use powder ensures trouble free application and the ultimate saving in lower reject rates is a most important economic factor.

Cleaning and Pretreatment

It must be emphasized that powder coating does not avoid the basic necessity to clean the surface of an article and a coating applied to a metallic surface which has been subjected to a controlled pretreatment will show improved corrosion resistance.

CLEANING

It may not be considered necessary to pretreat the metal surface but it is essential to clean it thoroughly to remove dirt, corrosion, scale, grease and oil so that the coating can adhere directly to the substrate.

The type and severity of contamination will largely decide the choice of cleaning material but it is likely that one or more of the following procedures will be effective as the first stage of surface preparation.

(a) Grit or Shot Blasting
(b) Solvent Wiping
(c) Vapour or liquid/vapour degreasing
(d) Emulsion cleaning by immersion or spray
(e) Alkali cleaning by immersion or spray
(f) Acid cleaning by immersion.

Mild steel is the most common surface for powder coating and the contaminants could consist of a mixture of mill scale, rust, oil and grease.

Mill scale is usually confined to heavy gauge plate and it serves as an excellent protection against corrosion before the plate is subjected to forming operations. Then the inherently brittle nature of mill scale results in it flaking from the metal surface which becomes an unsound base for subsequent coatings.

Removal of mill scale is therefore essential and heavy deposits of scale can only be removed by grit or shot blasting processes. The particle size of the grit or shot is important and the more coarse the grit the more effective is the removal of mill scale. However, a very coarse grit will produce a metal surface with high profiles which would not be the best surface for coating with a firmly adherent, corrosion resistant film. Fine grit is more expensive and takes longer to remove heavy deposits of mill scale. It does, however, provide a better surface for coating.

Shot is preferred to grit for it leaves a smoother surface.

Coating must take place immediately after grit or shot blasting because the cleaned surface is in a highly reactive state and rusting can quickly occur.

If oil and grease are also present on the surface to be prepared, the article must be degreased before shot blasting to prevent excessive contamination of the shot.

The main use for powder coatings on grit or shot blasted surfaces is in the coating of pipe.

A specification for "Chilled Iron Shot and Grit" is contained in BS 2451.

When the deposits of scale are light, they may be removed by the use of an acid cleaner to which may be added solvents and surfactants to remove oil and grease in one operation.

Both acid and alkali cleaners may be used on non-ferrous surfaces but care must be exercised with aluminium and zinc which can react vigorously with the cleaning chemicals. Expert advice is essential to decide the type, concentration and operating conditions for acid and alkali cleaners.

For removal of oil and grease only it may be possible to use a solvent wiping process which would also be effective in removing deposits of swarf and dirt.

Vapour degreasing with trichlorethylene, perchlorethylene or trichloroethane will remove oil and grease but not swarf and other solid contaminants and it would be necessary to wipe or spray water to rinse the articles and ensure freedom from solid deposits.

Liquid/vapour degreasing is more effective in the removal of solid deposits together with oil and grease.

The most common cleaning method (which is also the most economical) uses emulsified alkaline cleaners, and a wide range of these products are available.

Once again, the potential user is recommended to contact a specialist in this field who will advise the cleaning method best suited to his product.

A cleaned article must be subjected to thorough water rinsing before proceeding to pretreatment or coating, to ensure that no traces of the cleaning compounds are left on the metal surface.

PRETREATMENT

The surface of a metal is converted to an insoluble non-metallic coating by pretreatment and this thin film deposit improves the corrosion and chemical resistance of the coated article. In some instances it will also improve the adhesion of the powder coating but this will be, to some extent, dependent upon the thickness and granulometry of the deposited pretreatment layer.

Mild steel is generally pretreated by phosphating cleaned articles with a spray-applied solution or, alternatively, by immersion. The nature and thickness of the phosphate coating is controlled by the concentration and chemical composition of the solution, the operating temperature, the duration of spray or immersion and the presence of various additives which affect the

Fig. 148. View through pretreatment tunnel (Courtesy Controsion (Electrostatic) Ltd.)

film deposition. The solution can be acidic zinc phosphate which deposits insoluble zinc phosphate or, alternatively, an alkaline phosphate film. The zinc phosphate pretreatment invariably produces greater film thicknesses than deposited iron phosphate, and so is favoured for articles which will be subjected to the most severe corrosion conditions.

However, iron phosphate is less costly and so is preferred for many end uses.

Zinc coated metals have normally been pretreated by acidic zinc phosphate but, it has now been found that in practice, an alkaline phosphate pretreatment is adequate for most purposes. Zinc surfaces can also be treated by acidic chromate solutions to give an insoluble chromate film which retards the formation of "white rust".

Chromating is probably more effective than acidic zinc phosphate when there is any appreciable delay between pretreatment and coating but the adhesion and corrosion resistance properties of the film are inferior.

Chromate pretreatment for aluminium may result in the deposition of a yellow aluminium oxide chromate or a green aluminium oxide chromate dependent upon the choice of chemicals. These pretreatments are far more effective on aluminium than phosphate pretreatments which are sometimes employed when aluminium components are mixed with zinc and steel components.

EXISTING PLANT AND MACHINERY

The prospective user of powder coatings may already have a wet paint plant and he will obviously be concerned to utilize as much existing equipment as possible in setting up the new process.

A wet-paint spray booth is unlikely to be capable of adaptation to spraying powder coatings for it will lack the recovery system employed in powder booths which allows oversprayed powder to be collected and used again.

Wet-paint spray guns cannot be utilized nor can they be converted for powder application.

An existing good-quality air compression unit can be used in the new powder coating application plant provided that the expansion tank has an efficient oil and water trap, that a new air filter is installed to ensure that the supply of air is free from water and that the output from the compressor is sufficient. An air pressure of 60–80 p.s.i. should be adequate for most electrostatic powder spray units and the capacity of the pump to produce air must be related to the volume required when the new plant is fully operational. It is however more than likely that the compressor used with the wet paint plant will be suitable for the powder coating replacement unit.

The one item in existing plant which the end user will be most anxious to use, is the stoving oven.

Many industrial ovens are still functioning in a trouble free manner long after they have been completely written down in the Company accounts.

This does not mean, however, that the efficiency of the oven (after many years of continual use) has not deteriorated to the stage where it is more costly to continue with the old than to purchase a new oven. It is only just becoming painfully apparent to manufacturers that recent escalations in fuel costs require the most effective stoving equipment to obtain maximum thermal efficiency.

The capital expenditure on a new oven suitable for stoving powder coatings could soon be recovered by genuine savings in fuel costs. A wet paint oven requires a large throughput of air to convey solvent vapours away from the paint film and this results in very high heat losses. Powder coatings have little or no volatile by-products to be driven off during the stoving operation so that the more efficient hot-air-recirculation design is a feature of powder coating ovens.

Fig. 149. Square-tube table frames passing through wet paint booth before entering powder booth on system permitting either wet paint or powder application (Courtesy Controsion (Electrostatic) Ltd.)

Nevertheless, in spite of these cautionary comments against the continued use of an existing oven, it may still be cheaper to have this equipment modified to the improved air-recirculation system. While this is possible with gas or oil fired ovens it is doubtful whether an ancient infra red oven can be effectively converted for use with powder coatings. It should not be inferred from this remark that infra red ovens are unsuitable for use with powder coatings; indeed an example of this type of oven is given later in the chapter.

If the existing oven is incapable of stoving at a temperature higher than 140°C then it is unlikely to be suitable for stoving even the most rapid curing powder system. A powder coating is ideally cured by a gradual temperature rise up to 180°C or even 200°C so that the powder can be allowed to flow out into a continuous film before curing begins. An oven with a high temperature capability is not necessarily more expensive to operate than a low temperature oven, for the dwell time of the article in the oven is considerably shortened by the higher temperature stoving schedule. The capability to provide higher temperatures can also permit even shorter curing cycles if it is required to increase the track speed for increased production demand.

Powder coating may be introduced as a complementary method to existing wet paint systems and a prospective end-user will be concerned to know whether wet paint and powder coating can both be carried out simultaneously in the same area without detriment to either process. There are many instances where powder coating units have been installed alongside wet paint installations and, in the author's experience, this close proximity has not been detrimental to either process. However, the significant advantages to be gained with powder coating have been more apparent when compared side by side with wet paint processes and this has speeded the complete conversion to powder coating in many cases. An example of the side by side application of wet paint and powder is shown in fig. 149.

HAND OPERATED POWDER COATING UNITS

This equipment is relatively inexpensive and comprises a hand operated electrostatic spray gun, possibly a generator, a feed hopper, a spray booth and an overspray collection system.

The hand gun should be lightweight, robust and intrinsically safe in operation. Most hand guns weigh only 500 grams and therefore cause little discomfort to an operator using one continuously. Safe operation and robust construction are already specified in Britain by the B.A.S.E.E.F.A. regulations described in Chapter 11. The operator must not hold the gun with an insulative glove nor must he be wearing insulative footwear when he is operating the hand gun. He must at all times be earthed to prevent electrostatic build up.

One of the hand guns described in Chapter 10 operates without a high voltage generator by means of frictional charging within the barrel of the gun, but all the remaining hand guns are supplied by a generator. In some instances more than one hand gun can be connected to a single generator.

A feed hopper is standard equipment for all hand guns except laboratory models which have a feed cup attached to the gun. The hopper size and the method of conveying the powder varies with each manufacturer but the two methods commonly employed are fluidized air transport of powder or conveying by means of a worm screw.

The spray booth is the most important part of the equipment and there are a number of different design features that can bc incorporated. Electrostatic spray booths must be of metal construction to ensure dissipation of charge. If the booth is not used in conjunction with a conveyor system then a well earthed supporting hook must be provided to support the sprayed articles. A conveyor system must also be well earthed at the point of application.

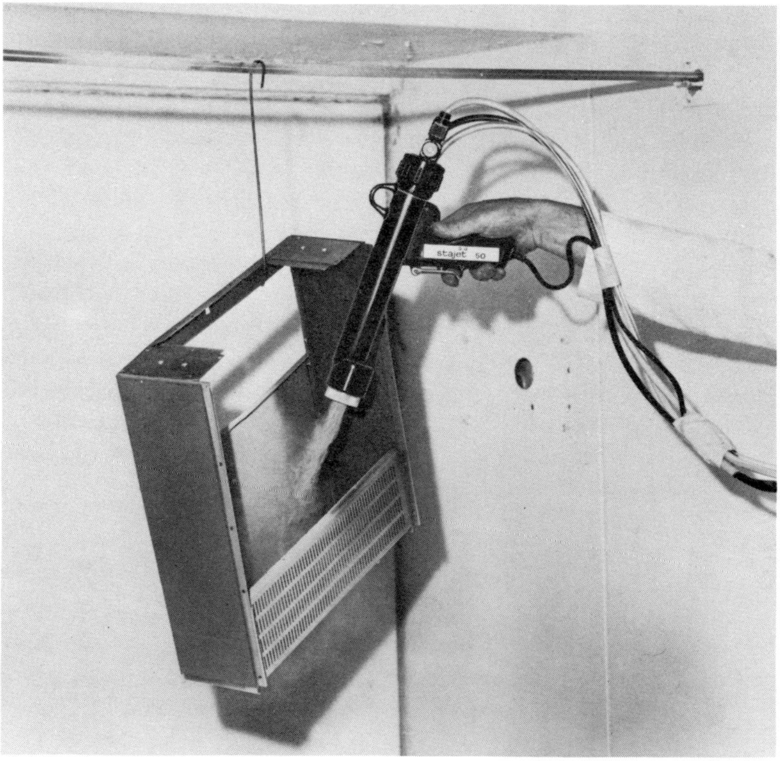

Fig. 150. Hand gun operation (Courtesy Binks Bullows Ltd.)

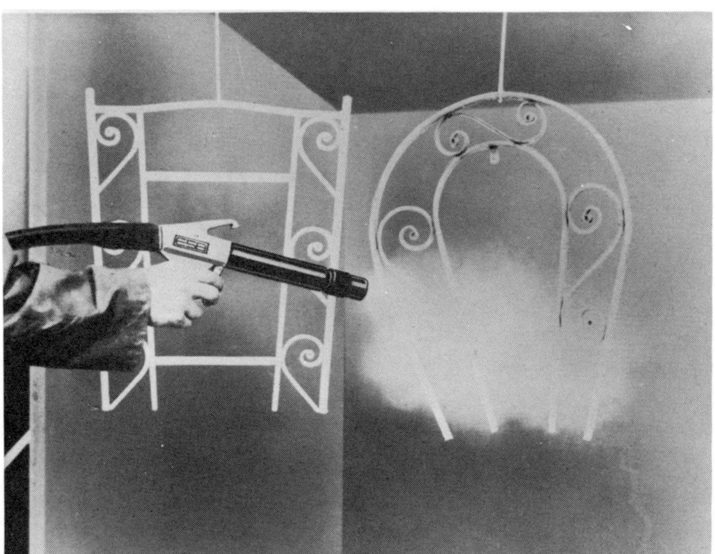

Fig. 151. Hand gun operation (Courtesy Spray Finishing Systems Ltd.)

The booth interior must be smooth to facilitate cleaning and have no recesses or ledges where powder can collect. Spraying may take place from one side or both sides. In the latter design the operator stations are staggered to prevent excessive charge build up close to the article.

The operator and the generator must be sited outside the booth.

When one grade of powder of one colour is sprayed continuously, the recovery system needs to be highly efficient to provide the best yield of oversprayed powder. This material can be used again but it is advisable to mix new and oversprayed powder to reduce differences in particle size distribution.

When purchasing a hand operated powder unit it is worth considering the possibility of converting this to a semi- or fully-automatic set-up at a later date. Certain makes of hand gun are capable of conversion and they could be mounted in a fixed position or even operated by a reciprocator.

Typical examples of hand gun application are illustrated in figs 150 and 151.

SEMI- AND FULLY-AUTOMATED PLANTS

The term semi-automated has not been well defined in the powder coatings application field but it can be used to describe an installation employing spray guns which operate automatically in a predetermined cycle but which are assisted by hand guns to "touch in" those areas which are missed by the automatic guns. It may also be used to describe a powder coating plant where guns operate automatically from a fixed position.

Fig. 152. The coating of column radiators (Courtesy Spray Finishing Systems Ltd.)

A semi-automated installation employing a hand gun (for filling in) would be less expensive for coating intricately shaped articles than an installation with a fully-automated programmed sequence. Fixed gun semi-automation is ideally suited for the coating of simple surfaces and this is illustrated in fig. 152 showing the coating of column radiators.

The term "fully-automated system" can describe one in which the continuous coating of simple articles by several guns mounted on reciprocators takes place. Such a system is shown in figs 153 and 154.

Fig. 153. Automatic powder application (Courtesy Binks Bullows Ltd.)

It is possible to programme the movement of the gun to follow the contours of the article and as many as sixteen different programmes can be selected at random to provide a versatile automatic spray system.

Automation of the powder plant is not confined to the spray operation, for it can also be applied to the powder recovery system with automatic re-cycling of overspray and reblending with fresh powder. Powder coatings may be stored in a silo connected directly to the reservoir of powder in the system and fed into the system automatically as the reservoir becomes depleted.

To limit the effects of fire occurring in the spray booth, the installation of heat sensitized monitoring devices, which can automatically close down operations and prevent excessive damage, is highly desirable.

A feature of the design of automatic powder spray booths is the provision of adequate ventilation to ensure that the concentration of powder is kept below the lower explosion limit. This concentration is in the range 0·02–0·04 ounces per cubic foot (20–40 grams per cubic metre). A powder usage of 20 kilograms per hour would therefore require that a minimum of 1,000 cubic metres of air per hour passed through the booth. This figure is usually

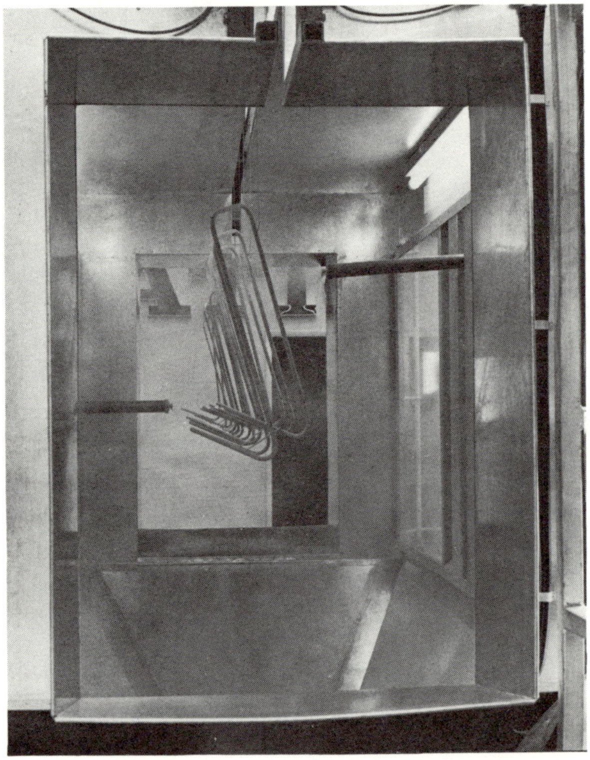

Fig. 154. Automatic powder application (Courtesy Volstatic Ltd.)

increased in practice by a safety factor of four to 4,000 cubic metres per hour or 66 cubic metres per minute. Assuming that the booth opening is 3 metres by 2 metres, the velocity of air at the booth face would be 11 metres per minute.

It is realistic to assume that of the powder sprayed, 60% will be deposited on the articles so that out of 20 kg sprayed 8 kilograms per hour will be conveyed to the recovery system at a concentration of 2 grams per cubic metre.

A filter cloth collection system will have a dust extraction efficiency as high as 99·5% so that the air exhausted from the collector would have a maximum contamination level of 10 milligrams per cubic metre. It should therefore be possible to exhaust this air to atmosphere or return it directly to the workshop.

Cyclone recovery systems are unlikely to be more than 80% efficient at these low particle sizes and they must therefore be used in conjunction with a cloth filter.

The minimum ignition energy required to initiate an explosion, if the lower level of explosivity is exceeded, is in the region of 25–100 millijoules. This energy level is approximately 100 times higher than the energy required to initiate an explosion with a solvent vapour/air mixture also at the lower limit of explosivity.

Colour Changing

A distinct advantage of wet paint systems is the ease with which colour changes can be effected. The paint lines can be quickly purged when the new colour is introduced and there is little or no contamination. A similar procedure with powder coatings would result in a speckled finish and this defect would be apparent for a considerable period after the colour change.

The task of cleaning down the guns, booth and recovery system would occupy the operator for at least one hour and this would not be acceptable in a production schedule requiring frequent changes of colour.

Cleaning procedures have been recommended for some types of powder application equipment in which this time can be reduced, but for many large potential users, the need to be able to change colours as rapidly as in wet paint processes, puts the use of powder out of court for the present time.

Colour changing is now accepted by equipment manufacturers as a basic problem and considerable effort is being devoted to solving this issue to meet the needs of the automotive industry. To change colour quickly with powder coatings at present, it is necessary to provide a separate powder coating booth for each colour, with a separate recovery and regeneration system, and separate guns and feed hoppers. Along these lines several systems have been proposed for quick colour change and some of them are currently in use in industrial premises. For example:

In 1973 a fully-automated electronically-controlled eight colour plant was installed in a plant belonging to a company manufacturing do-it-yourself appliances and with it, it has proved possible to change colour in 47 seconds. The eight separate units are sited close to the product conveyor. When a colour change is required several jigs are removed from the conveyor track and a colour code flag is inserted in the position vacated by the first jig. This device alerts the control circuit to clear away the existing colour plant which is withdrawn from its position beneath the conveyor and replaced by the unit supplying the new colour that has been selected. The operation is fully automatic and has a "fail-safe" capability to ensure that each stage is performing correctly.

Although the cost of the installation was high, this cost is being handsomely repayed by the improved efficiency of painting.

A much less costly version of this technique could be applied to a two colour plant with manual selection at the central control cabinet. The unit withdrawn from use could be cleaned down to use with any colour while the replacing unit continued to function. Theoretically, this "two colour" set up could be extended to apply a number of different colours without interrupting production time.

Future developments will be directed towards much greater deposition efficiency of powder coatings so that recovery and recirculation are unnecessary. It is envisaged that the requirements of the automotive industry for a variety of body colours could be met by quick change of colour feed to one gun, the gun being purged by a small amount of new powder which could be reprocessed for other end-use applications.

Jigging and the Cleaning of Jigs

The design of the jig is an important aspect of successful powder coating and the shape and size will generally be decided after the initial application trials in the test centre of the equipment supplier. A multigun installation will apply powder more efficiently than a single gun and it will be possible to carry many articles on one jig providing they are relatively small in size.

It is essential that the jig at all times makes good electrical contact with the article and with the earthed support in the booth or on the earthed conveyor track.

If good electrical contact is not maintained the efficiency of application will be reduced and charged powder will be deposited in increasing quantities on the walls of the booth and the conveyor track.

The jig will not have to be cleaned after every coating operation but it is unlikely to remain an effective conductor after three coating sequences. It will be almost impossible to prevent the contact point between the jig and the article from receiving an insulative coating but attention should be given to the point of contact with the conveyor track. If the jig can be slotted into a recess in the top of the supporting bar attached to the conveyor chain so that there is no clearance for the penetration of powder, it will not be necessary to clean this contact point very frequently.

Cleaning jigs is difficult for the majority of powder coatings are based upon thermoset epoxy resins. These are impervious, after stoving, to strong organic solvents and they are relatively unaffected by immersion in hot strong alkali solutions. The only effective stripping solutions are those based upon cresylic acids and glacial acetic acid which are most unpleasant to use. If it is decided to use this type of stripping solution then its surface must be covered at all times by a layer of water and the operator must use neoprene gloves and wear protective clothing with effective goggles to protect the eyes.

One alternative method for cleaning the jigs is to burn off the stoved film at temperatures above 400°C; another is to remove the coating by grit blasting.

Perhaps the easiest method is to design the jigs with a replaceable contact point which slots into the supporting arm and this could be removed after each trip and replaced with a clean metal contact. This would avoid the necessity to clean the whole jig except at routine maintenance times.

Recovery Systems

These are generally designed by the equipment manufacturer and form an integral part of the spray unit. The type of dust collection unit employed for powder recovery has already been described in Chapter 11 and there is no need to enlarge on the previous comments.

Fig. 155. Large recovery unit (Courtesy Tycoate Industrial Finishes)

A large automatic powder application plant will have a recovery system which returns the collected oversprayed powder to the feed reservoir, after sieving to remove fibres and large agglomerates accumulated on route. This cycle can be operated in a continuous or intermittent sequence which will depend largely on the quantities of overspray powder to be recycled. Smaller units will only be supplied with a dust collector and it will be necessary for the operator to return the overspray and resieve it before blending with fresh powder in the feed hopper. With this arrangement it is essential to purchase a separate sieve for it would be unwise to return the oversprayed powder without removing fibrous contaminants. An example is shown in fig. 155 of a large, recycling recovery unit where the cyclone in the foreground returns the overspray to the Ofner collector where it is returned via a cylindrical sieve to the feed hopper.

In the foreground of fig. 156 a standard dust collection unit is shown with the catchment pot for recovering oversprayed powder.

Fig. 156. Dust collection unit for powder recovery

OVENS

Radiant Heat Ovens

Modern stoving ovens, fired by gas, employ radiant heaters at 800°C.

Heating by radiant energy is more efficient than convection heating in that energy is transferred into the article much more quickly. Hence stoving times of between 3 and 5 minutes are possible. It is however difficult to assess the actual temperature of the stoved articles as they pass through the radiant heat oven and the correct oven settings are often obtained by trial and error methods.

Since the temperature obtained by an article in a short time will depend on its mass, a radiant heat oven will not give consistent cure where articles of heavier gauge are included with light gauge ones in a production run. Several well-known manufacturers are curing powder coated articles of uniform size and thickness successfully by radiant heat methods.

The radiant heat oven is usually less costly to purchase, occupies much less space, heats up to temperature more rapidly than a convection oven and, it is claimed by one major manufacturer of radiant heat ovens, that they are no more costly to operate.

Convection Ovens

The main argument in favour of convection ovens is the even distribution of air temperature that can be obtained throughout the oven, thereby ensuring

Fig. 157. Radiant heat oven (Courtesy Maywick Ltd.)

that the articles are cured correctly regardless of shape, size and their position within the oven. Heat reaching the articles passing through a radiant heat oven will vary as the inverse square of the distance from the emitter surface so that an area on the article sited 4 in. from the emitter will only receive one quarter of the heat reaching an area 2 in. from the heat source.

Gas is the most reliable method of heating a forced air-circulation powder oven but care should be taken to avoid a gas supply containing traces of acidic oxides for these are likely to discolour white powder coatings cured with amine hardeners. Traces of trichloroethylene will also cause a reduction in the gloss level of cured epoxy powder coatings and the oven should be sited well away from degreasing plants.

A typical forced air-circulation oven is shown in fig. 158 comprising a dry-

Fig. 158. Overhead air circulation oven (Courtesy Controsion (Electrostatic) Ltd.)

off section on the left-hand side where the articles have one pass and, after coating, cure is effected by three passes along the right hand side of the oven.

In conclusion and for completeness it is necessary to discuss briefly some of the major problems encountered in the industry.

PROBLEMS ENCOUNTERED IN POWDER COATING UNITS

Contaminated or Sub-standard Powder

This may be applied to articles with the result that surface defects are apparent in the stoved film. Unless the powder is very old stock it is unlikely to have developed these faults in storage. Most likely causes are inadequate pre-treatment or poor cleaning, silicone grease contamination during servicing of the conveyor track, or the mixing of two incompatible powders. It should not be assumed that epoxy powders from two different suppliers are compatible in all proportions and this aspect should be tested out before purchasing from a new supplier. Small traces of polyester and acrylic powders may produce severe pinholing problems with epoxies and the applicator should avoid using chemically different types of powder coating in the same booth.

When the fault has been located and the plant is again running normally, it will be necessary to recoat the spoiled articles. If the surface defects are severe it will be wise to completely remove the stoved film. Slight defects, however, including the circumstance when the powder is of good quality but off shade, can be remedied by light sanding to abrade the surface, followed by heating the article and recoating while it is still warm.

Brittle Powder Film

If the powder film is found to be excessively brittle, it is invariably due to insufficient curing because powder coatings are not adversely embrittled by overstoving. Poor resistance to acetone or methylethylketone will confirm that the film is undercured. Restoving the article at the correct curing schedule is unlikely to complete the film cure, and the safest remedy is to remove the faulty coating and recoat the article.

Blow Holes

Porous castings develop blow holes when coated with powder coatings. Air entrapment in the pores of the metal will disrupt the powder coating after flow has commenced. The only way to overcome this fault is to prime the castings with an air drying primer specially designed for this purpose.

Poor Matt Finish

A matt finish is found to be too high in gloss. This is again a fault caused by under stoving and can be resolved by raising the oven temperature.

15

Future Developments

SUMMARY

This final chapter is itself a summary of the previous chapters, with speculative comments on future development of powder coatings. The technology of these products has reached the stage where innovations are a daily occurrence and the scope of these materials in the industrial coatings field will continue to expand with improved formulations, stricter control of particle size and shape, easier methods of application and handling, with an efficiency of powder utilization on to the finished article which cannot be matched by liquid coatings. Above all their economy in the use of petroleum products and their elimination of pollution will be maintained.

THE FUTURE MARKETS FOR POWDER COATINGS

Without doubt the most significant development in powder coatings is "thin film" technology which will enable powder coatings to become economically competitive with wet paint processes. Increasing costs of petroleum based ingredients will equally affect the price of resins used in both wet paint and powder coatings, but the cost of solvents in the former, solvents which ultimately go to waste, will assume an even greater importance as prices rise and tip the balance more firmly still in favour of powder coatings.

It is not possible, however, to equate the cost saving of the thin film coating directly to the thickness of the deposited film, for the product will have to be adjusted in pigment to binder ratio to provide improved opacity at the lower film weight.

Although the cost of titanium dioxide is less than that of resin, an increase in titanium dioxide content will raise the specific gravity of the powder and reduce the square metres covered per kilo of powder. It will also be more difficult to disperse the higher concentration of pigment and this will add to production costs.

A major contribution to the cost of powder coatings at present is the high level of capital expenditure required for the manufacturing process.

Alternative methods of manufacture suggested to date seem likely to increase this capital outlay, but it must be borne in mind that no piece of equipment has yet been designed specifically for the production of thermosetting powder coatings. As the compatibility of the resins and hardeners is improved, less work will be necessary to compound the finished product and the high shear forces required to manipulate thermoplastics will not be needed in the production of the lower molecular weight thermoset powder coatings.

An expanding market for powders as a replacement for industrial stoving paints is easy to envisage, and it will appear as the quality of powders is improved.

There are, however, many possible outlets for powders coatings which could only be satisfied by wet paints if they were applied by costly application procedures and techniques which normally result in a high reject rate.

Powder coatings are likely to be investigated as an alternative to vitreous enamel in the domestic appliance field, as base coats for the vacuum deposition of metals, as coatings for glass containers and safety glass laminations, as a decorative finish for compounded board, in badge manufacture and as finish coatings for a range of plastics. The use of powders will also increase considerably in the electrical industry for example for uprating the power output of the coated component.

There are many more possibilities for powder coatings because they offer improved adhesion and better film performance when applied to difficult substrates; another most important factor in their favour is the lower rate of rejects caused by faulty application.

FUTURE RESIN DEVELOPMENTS

Within a decade there will be significant developments in epoxy resins to improve their exterior durability and their resistances to heat and light. To achieve these qualities it will be necessary to remove aromatic groups from within the epoxy resin molecule. Already this has been attempted by hydrogenation, but the cost of the process and the selection of the correct catalyst has so far made it commercially untenable. It is likely that the complete removal of the aromatic nucleus would result in a marked reduction of the chemical resistance properties of the epoxy resin, so that a compromise may have to be reached between the two extremes, whereby the optimum development of all desirable properties is attained.

Other improvements in the manufacture of epoxy resins will result in closer control of the molecular weight spread, smoother cross-linking reactions due to designed intermolecular branching, and an elevation of the glass transition temperature.

It will be much more difficult to develop new polyester resins for these are manufactured at higher temperatures than are epoxy resins by a process in which esterification is competing with etherification and polymerization. Nevertheless, it should be possible to control the molecular weight distribution by the use of symmetrical polyhydric alcohols with primary hydroxyl functionality and dibasic acids, which do not promote the cyclization of low molecular weight fractions. Greater use will be made of synthetic fatty acids, and while these costly components will make the polyester resins much more expensive than the alkyd resins used in wet paint, they will provide powder coatings of performance superior to the even more costly acrylics.

Polyurethane powder coatings are unlikely to become universally acceptable due to the high cost of raw materials, and the loss in coating weight of the stoved film, by at least 5%, will add to the cost problems with increased disapproval by conservationists.

There is already considerable interest in blends of epoxy with polyester resins, and this work will be extended to include blends with acrylic polymers, so that mixtures of the various resin types are more likely to lead to early successes than are refinements to the basic resins alone.

It is difficult to foresee how the future development of acrylic powder-coating resin will go in view of present problems with the control of molecular weight distribution during emulsion polymerization. One possible approach is the preparation of acrylic resins by solution polymerization, followed by compounding with pigments, curing agents and flow additives in conventional wet-paint dispersion equipment. The paint could then be converted to a powder, with controlled particle size, by the use of spray drying techniques.

Powder coatings based upon thermosetting vinyl resins have already been developed in the United States and this type of product will become more readily available as the problems of manufacturing vinyl powder coatings, of the correct particle size, are solved.

Phenolic resins may be adapted for use in powder coatings, either as the sole film-forming resin, or in conjunction with epoxy resins. Flexible compositions of phenolic resins could be usefully applied for the lining of containers or tubes where excellent water resistance is required.

FUTURE DEVELOPMENTS OF CURING AGENTS

The curing agents used for cross-linking epoxy resins are currently based upon catalysed dicyandiamide, substituted dicyandiamide or alicylic amine compounds.

These have the disadvantages of poor solubility, poor compatibility with the solid grade epoxies, and melting points which are considerably higher than the processing temperatures used for the compounding of powder coatings. When the amine compound is soluble and compatible, and low enough in melting point to be capable of homogeneous dispersion in the epoxy resin, it is too reactive for hot melt compounding and in consequence the powder coating has poor flow and reduced storage stability. Amine hardeners also exhibit certain side effects in the cured coating such as film discoloration and embrittlement upon ageing.

The alternative to amine curing is the anhydride cure, but again many of these compounds are neither soluble nor compatible and those suitable for use as hardeners possess high melting points. Additional drawbacks to the anhydride curing agents are in the high proportion needed for crosslinking the epoxy resin, their irritant nature and their proneness to be hygroscopic. However, in spite of their adverse characteristics, anhydride hardeners produce properties in the cured film which are better than those afforded by the amines.

A pointer to future trends in hardener development is given by the success of the polyester epoxy system, where the epoxy is cross-linked by the high acidity of the polyester. This resinous hardener has the obvious advantages of comparable melting range to the epoxy resin, complete compatibility, low reactivity at the compounding temperature, good melting properties to assist the incorporation of pigments and smoother reaction at the curing temperature due to a reduction in steric hindrance.

Crystalline hardeners of high reactivity will eventually be replaced by resinous hardeners to give the powder coating formulator the versatility of the paint chemist.

THE FUTURE DEVELOPMENTS IN POWDER COATING FORMULATIONS

At present, the powder coating formulator is restricted by the type of resin and hardener systems available, but this situation is likely to be improved considerably in the near future. If manufacturing techniques are not altered drastically, it will be necessary for the formulator to investigate the use of additives to improve the compounding efficiency of the extruders.

Agents will need to be assessed for improving pigment wetting, for reducing surface tension during the melting of the powder particles, and for improving film flow and adhesion. As more information is gained about the electrical resistivity of powder particles, it may be important to control this property with additives similar in chemical composition to the wetting agents.

It is fairly certain that adequate dispersion in a mix of high proportions of titanium dioxide will not be achieved by a single pass through an extruder, even with the aid of a wetting agent, and the formulator will have to investigate methods for predispersing pigments under high shear to the point where the originally agglomerated pigment particles can then be incorporated in a single extrusion operation. This masterbatch technique will not be confined to titanium dioxide, for strong coloured pigments must also be predispersed to enable full development of colour strength. High quality dispersion of carbon black will require the application of techniques which have long been considered to be essential in the plastics compounding field, where much lower concentrations of black are used to colour the finished product.

The formulator will also devote considerable effort to the assessment of suitable flow control agents which may be closely tied in with the work on wetting agents.

FUTURE DEVELOPMENTS IN POWDER COATING MANUFACTURE

The techniques for manufacturing powder coatings are likely to undergo the most radical changes in the future development of more sophisticated products.

The hot melt methods now favoured are over-elaborate because they undertake the task of homogenizing components which are not strictly compatible and, in fact, each stage of the production process is an attempt to improve the homogeneity of a system which is basically heterogeneous. Improvements in resins, curing agents and predispersion techniques will make it unnecessary to employ the high shear forces operating in present day extrusion processes.

The lower reactivity of the resinous hardener will permit an increase in dwell time in the heated zone of the compounding machine, with improved dispersion at lower work levels. An alternative to the hot melt extrusion process, which has been widely publicized in recent technical literature, is the spray drying technique, where the components are initially dispersed in a carrier solvent by conventional wet paint dispersion methods, followed by the complete removal of the carrier solvent when the liquid is sprayed as a jet of finely divided droplets into a heated vessel under high vacuum.

However, this method, which is no doubt, highly successful for the large scale manufacture of dried milk or powdered soup, would be extremely difficult and costly to apply to the handling of paint products, and it is unlikely to progress beyond the academic stage (except for acrylics) as a method for powder coating manufacture.

It is much more probable that water will be used as the intrinsically safe, cheap carrier for powder compounding and fine grinding. If the use of water could be applied now to existing production methods there would be immediate and obvious advantages.

Premixing would be dust-free and could be accomplished as a slurry in an open high speed mixer. The slurry could be metered into the extruder together with steam to assist the melting of the resinous components. Low viscosity extrudate might then be precipitated, finely divided solid could be filtered to remove the excess water and dried on a continuous band. Powder coating production by this method would be completely free from the risk of fire or explosion and there would be no health hazard for the operator. The lubricating action of the water carrier would promote the formation of spherical particles and a narrow spread of particle size. There is also a distinct possibility that pigment agglomerates could be easily dispersed in an aqueous carrier and facilitate the compounding process.

A less speculative development in powder manufacture is the method of fine grinding by fluid energy mill to give smaller particles with a much narrower distribution of size. At present it is difficult to apply these fine powder coatings for they are not easily fluidized and they are not deposited as readily as larger particles when applied electrostatically.

However, fine powders will give lower film thickness and reduce the problem of air entrapment as a direct result of the lower voidage between the particles.

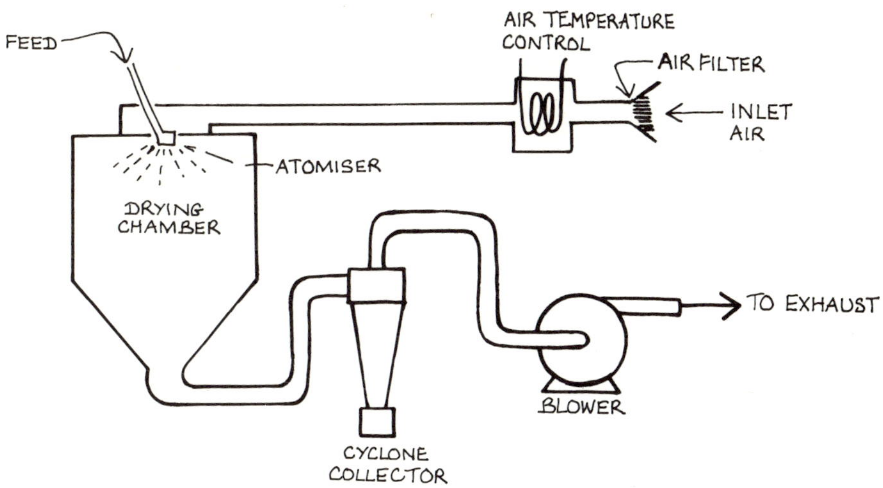

Fig. 159. Schematic drawing of spray drying process

Improved grinding efficiency will be obtained in conventional pin disc mills by application of freeze grinding techniques, and there is no doubt that this procedure will be widely applied in powder coating manufacture for the fine grinding of polyesters and acrylics at temperatures well below their glass transition temperatures.

FUTURE DEVELOPMENTS IN POWDER APPLICATION

Developments in fluidized bed application of powder coatings will be confined to minor innovations designed to improve the versatility of this coating method.

The patent literature is already considerable and there are few everyday articles which have not been investigated for coating by fluidized bed. The major advantage of fluidized bed coating is the low cost of the capital equipment compared with electrostatic spray, and it is likely that further attempts will be made to effect a compromise between the two processes, fluidised bed and electrostatic spray. The compromise has not been achieved with the electrostatic fluidized bed in its present form. It is possible that the "cloud chamber" process could have been a midway stage between the two, but there were many problems arising from the excessive volume of powder used in relation to the volume of the articles passing through the cloud. The risk involved in the use of powder concentrations in excess of the lower limit of explosivity, was a further condemnation of this process.

It is the field of electrostatic spray equipment which will see the most significant improvements, with considerable efforts being made to resolve certain basic problems. These include changing quickly from one colour, or type of powder, to another, without contamination between the products. This will involve new designs for the booth and collection units, and simple cleaning processes for the gun, powder feed line and feed hopper. There are many powder coating equipment manufacturers who believe that the real answer to this problem is to improve deposition efficiency to the level where powder recovery is unimportant, but they rightly point out that they will never achieve the efficiency of deposition required by automotive manufacturers, without the full co-operation of the powder producers. It will be necessary to manufacture powders of precise particle size, shape and electrical resistivity, to allow equipment manufacturers to devise new methods for charging the particles and conveying them uniformly to the article being coated.

Perhaps it will be possible to get away from the closed booth and use electrostatic fields to precipitate the small amount of overspray, which would remove the need for excessive air ventilation and avoid the use of cyclone or bag collection units.

CONCLUSION

It is difficult to speculate about the important future developments in powder coatings, for it is only adding further controversy to a new field of coatings technology which already has an exciting future, commercially and technically.

Appendix 1

LIST OF SUPPLIERS & SERVICES

Epoxy Resins

CdF Chimie
Ciba-Geigy
Dow Chemical
Emser Werke
Schering
Shell Chemicals

Curing Agents

Bayer
Cargill
Ciba-Geigy
Dow Chemical
Emser Werke
Schering
Shell Chemical
Veba Chemie

Inorganic Pigments

Blythe Colours
Cowan
ICI
Reckitts
SCC Colours

Drum Hoop Mixers

J. Engelsmann
KEK
Winkworth

Conical Mixers

Nautamix

Z Blade Mixers

Baker Perkins
Beken
Winkworth

Polyester Resins

BASF
Bayer
BIP
CdF Chimie
Dynamit Nobel
Scado
UCB
Veba Chemie

Flow Additives

BASF
Cole Chemicals
Eastman Chemical
Hoechst Chemical
Monsanto
Scado

Organic Pigments

BASF
Bayer
Ciba-Geigy
Hoechst.
ICI
Sandoz

Cone Blenders

Baker Perkins
Gardners

High Speed Blenders

Baker Perkins
Dierks (Diosna)

Extruders

Baker Perkins
Buss
Eickhoff
Werner & Pfleiderer

Acrylic Resins

BASF
Bayer
Cole Chemicals
UCB

Titanium Dioxide

Kronos
Laporte
Sachtleben
Tioxide International

Carbon Black

Cabot Carbon
Columbian International

Metallic Pigments

Alcoa
British Aluminium
English Metal Powders
S. Fry
Johnson & Bloy
Wolstenholme

Horizontal Mixers

Gardners
Lodige
Winkworth

Air Jet Mixers

Gardners

Cooling Bands

Baker Perkins
Buss
Eickhoff
Sandvik
Werner & Pfleiderer

Vibratory Feeders

KEK
Nautamix

Classifiers

Alpine
British Rema
Hosokawa
Mikropul-Ducon

Fluidized Bed.

TAM-SAMES
Volstatic

Pretreatment Plant

ICI
Pyrene

Nylon Powders

ATO Chimie

Test Stations

BAM
BASEEFA
CERCHAR
Fire Research Station

Granulators

Alpine
KEK
USI

Sieves

Alpine
Drury Powders
Hosokawa
KEK
Russell Finex

Electrostatic Fluidized Bed

TAM-SAMES

Ovens

Stein Atkinson Stordy
Maywick

Cryogenic Grinding

Air Products
British Oxygen

Fine Grinding Machinery

Alpine
Hosokawa
KEK
Mikropul-Ducon

Dust Collection Units

Alpine
British Rema
Drury Powders
Dust Control Equipment
KEK
Mikropul-Ducon

Electrostatic Spray Equipment

Ashdee
Controsion-Aerostyle
DeVilbiss
Drury Powder
Estey
Gema
Kremlin
Mindon
Nordson
Ransburg-Peabody
Spray Finishing Systems
TAM-SAMES
Volstatic

ADDRESSES OF SUPPLIERS AND SERVICES
including the Selling Agents in the U.K.

ATO CHIMIE, Tour Aquitaine, 92080—Paris—La Defense (France) Tel. 256.61.61. Telex 62953.
Technical Centre-Coatings, Kiln Road, Shaw, Newbury, Berks. Tel: Newbury 5048/9. Telex 84689

ALCOA OF GREAT BRITAIN LTD., Alcoa House, PO Box 15, Droitwich, Worcestershire. Tel: 090-57 3411. Telex 338518

ALPINE AG, 89 Augsburg 1, Postfach 101109, Federal Republic of Germany. Tel: 0821 94011. Telex 053802
M & M Process Equipment Ltd., Frogmore Road, Hemel Hempstead, Hertfordshire. Tel: Hemel Hempstead 63621.

AIR PRODUCTS LTD., St. Georges Square, New Malden, Surrey.

ASHDEE, PO Box 325, 10 South Tenth Avenue, Evansville, Indiana 47701, U.S.A.

BAKER PERKINS CHEMICAL MACHINERY LTD., Westfield Road, Peterborough PE3 6TA. Tel: 0733 262000. Telex 32185.

BUNDESANSTALT FUR MATERIAL-PRUFUNG, B.A.M., Berlin, Federal Republic of Germany.

BASEEFA, Harpur Hill, Buxton, Derbyshire.

BASF (UK) LTD, PO Box No 4, Earl Road, Cheadle Hume, Cheadle, Cheshire SK8 6QG. Tel: 061-485 6222. Telex 668682.

BAYER (UK) LTD., Bayer House, Richmond, Surrey TW9 1SJ. Tel: 01-940 6077. Telex 927027/8.

BEKEN ENGINEERING LTD., Ripple Road, Barking, Essex. Tel: 01-592 2227.

BLYTHE COLOURS LTD., Cresswell, Stoke-on-Trent, Staffordshire. Tel: 0782 85666. Telex: 36115.

BRITISH ALUMINIUM CO. LTD., PO Box 108, Latchford Locks Works, Warrington, Lancs. WA4 1NP. Tel: 0925 35314. Telex 62652.

BRITISH INDUSTRIAL PLASTICS LTD., Chemicals Division, Popes Lane, PO Box 6, Oldbury, Warley, West Midlands B69 4PD. Tel: 021-552 1551. Telex 337261.

BRITISH OXYGEN CHEMICALS LTD., Vigo Lane, Chester-le-Street, Co. Durham. Tel: 089-424 2631. Telex 53163.

BRITISH 'REMA' MANUFACTURING CO. LTD., PO Box 31, Imperial Steel Works, Sheffield, Yorks S9 5GA. Tel: 0742 49054. Telex 54115.

BUSS LTD., Basle/Pratteln, Switzerland.
Buss-Hamilton Ltd., Clifton House, 83/89, Uxbridge Road, London W5 5TG. Tel: 01-579 7381/2. Telex 25613.

CABOT CARBON LTD., Stanlow, Ellesmere Port, Cheshire L65 4HT. Tel: 051-355 3677. Telex 62261.

CARGILL INC., Minneapolis, U.S.A.
Victor Blagden & Co. Ltd., A.M.P. House, Dingwall Road, Croydon, Surrey CR9 3QU. Tel: 01-681 2341.

CdF CHIMIE, Tour Aurore, Cedex No. 5—92080 Paris—La Defense, France. Tel: 788.33.11. Telex CDF CH D 61.826 F.

CERCHAR-Industrie, Laboratoire du Centre d'Etudes et de Recherches des Charbonnages de France, Verneuil-en-Halatte, France.

CIBA-GEIGY (UK) LTD.,
Pigments Division: Wythenshawe, Manchester M23 9ND. Tel: 061-998 7121. Telex 668083.
Plastics Division: Duxford, Cambridge CB2 4QA. Tel: Sawston 2121. Telex 81101.

COLE CHEMICALS LTD., 7–15, Lansdowne Road, Croydon CR9 2HB. Tel. 01-686 4411. Telex 23864.

COLUMBIAN INTERNATIONAL (G.B.) LTD., Annabelle House, 28 Staines Road, Hounslow, Middlesex. Tel: 01-570 3296. Telex 263614.

CONTROSION ELECTROSTATIC DIVISION, Aerostyle Ltd., 13 Sunbeam Road, North Acton, London. NW10 6JT. Tel: 01-965 3464. Telex 932627.

COWAN BROS (STRATFORD) LTD., Barbers Road, Stratford, London, E15 Tel: 01-534 3311.

DEVILBISS CO. LTD., Ringwood Road, Bournemouth. Tel: 020-16 3131. Telex 41213.

DIOSNA—Dierks & Sohne, Osnabruck.
Buss-Hamilton Ltd., Clifton House, 83/89, Uxbridge Road, London W5 5TG. Tel: 01-579 7381/2. Telex 25613.

DOW CHEMICAL LTD., Heathrow House, Bath Road, Hounslow TW5 9QY. Tel: 01-759 2600. Telex 934626.

DRURY POWDER PLANT LTD., Charles Henry Street, Birmingham B12 0SJ. Tel: 021-622 4147/8.

DUST CONTROL EQUIPMENT LTD., Thurmaston, Leicester LE4 8HP. Tel: Syston 3333. Telex 34500.

DYNAMIT NOBEL (UK) LTD., Greenco House, 36–40, High Street, Slough, Bucks. SL1 1HF. Tel: Slough 35422. Telex 848243.

EASTMAN CHEMICAL INTERNATIONAL A.G., PO Box 66, Kodak House, Station Road, Hemel Hempstead, Hertfordshire. Tel: Hemel Hempstead 62441. Telex 825227.

GEBR. EICKHOFF, D-4630 Bochum, Hunscheidtstrasse 176, Postfach 629, Federal Republic of Germany. Tel: 3 05–1. Telex 08 25 804.
Calmore Machinery Co. Ltd., 31, Fritham Close, Totton, Nr. Southampton, Hampshire. Tel: 04216 3738. Telex 47388.

EMSER WERKE AG., Domat/Ems, Switzerland.

J. ENGELSMANN A.G., Ludwigshafen/Rh 3, Federal Republic of Germany.
Calmore Machinery Co. Ltd., 31 Fritham Close, Totton, Nr. Southampton, Hampshire. Tel: 04216 3738. Telex 47388.

ENGLISH METAL POWDERS LTD., PO Box 4, Widnes, Cheshire WA8 0PC. Tel: 051-424 2611. Telex 62171.

ESTEY DYNAMICS CORPORATION, Red Bank, New Jersey, U.S.A.
Electropaint Ltd., Shire House, Buckingham Road, Lichfield, Staffordshire.

FIRE RESEARCH STATION, Borehamwood, Hertfordshire WD6 2BL. Tel: 01-953 6177.

FRY, S & CO LTD., 23 Benwell Road, London N7 7BN. Tel: 01-607 6789. Telex 22821.

GARDNERS OF GLOUCESTER LTD., Bristol Road, Gloucester GL1 5RP. Tel: 0452 24087. Telex 43179.

GEMA LTD., Kunklerstrasse 9, CH-9015, St. Gallen, Switzerland. Tel: 071 311313. Telex 77402.
Tycoate International Ltd., Gardiners Place, West Gillibrands, Skelmersdale, Lancashire WN8 9SP. Tel: 0695 22171. Telex 627775.

HOECHST CHEMICALS LTD., Hoechst House, Salisbury Road, Hounslow, Middlesex. Tel: 01-570 7712. Telex 23284.

HOSOKAWA EUROPE LTD., St. John's Estate, Tylers Green, High Wycombe, Buckinghamshire. Tel: 049-481 2824. Telex 837208.

IMPERIAL CHEMICAL INDUSTRIES LTD., Imperial Chemical House, Millbank, London SW1. Tel: 01-834 4444. Telex 32124.

JOHNSON & BLOY LTD., 'Metana Works' Priestley Way, Crawley, Sussex. Tel: Crawley 23152. Telex 97504.

KEK LTD., Hulley Road, Hurdsfield Industrial Estate, Macclesfield, Cheshire. Tel: Macclesfield 28733. Telex 668623.

KREMLIN SPRAY PAINTING EQUIPMENT LTD., 839 Yeovil Road, Slough, Buckinghamshire SL1 4JA. Tel: Slough 30038. Telex 847969.

KRONOS TITANIUM PIGMENTS LTD., St. Ann's House, Parsonage Green, Wilmslow, Cheshire SK9 1HG. Tel: 099-64 29511. Telex 669055.

LAPORTE INDUSTRIES LTD., PO Box 26, Grimsby, Lincolnshire. Tel: 0469-23171. Telex 52173.

GEBR. LODIGE MASCHINENBAU-GMBH, 479 Paderborn, Elsener Strasse 7/9, Postfach 729, Federal Republic of Germany. Tel: 3471/72. Telex 0932269.

MAYWICK (HANNINGFIELD) LTD., Rettendon Common, Chelmsford, Essex CM3 5HY. Tel: 0245 400637.

MIKROPUL-DUCON LTD., Towerfield Industrial Estate, Shoeburyness, Essex SS3 9QU. Tel: Shoeburyness 3511. Telex 99346.

MINDON ENGINEERING (NOTTINGHAM) LTD., Brookhill Industrial Estate, Station Road, Pinxton, Nottingham NG16 6NS. Tel: Ripley 810034.

MONSANTO, Monsanto House, 10–18, Victoria Street, London SW1H 0NQ. Tel: 01-222 5678. Telex 916093.
Durham Raw Materials Ltd., 1–4, Great Tower Street, London EC3. Tel: 01-626 4333.

NAUTAMIX UK LTD., National Westminster House, 2 Chestergate, Macclesfield SK11 6BA. Tel: Macclesfield 28833.

NORDSON (UK) LTD., Wenman Road, Thame Industrial Estate, Thame, Oxfordshire. Tel: Thame 3171. Telex 837214

PYRENE CHEMICAL SERVICES LTD., Ridgeway, Iver, Buckinghamshire. Tel: Iver 1812. Telex 84693.

RANSBURG PEABODY LTD., Weybridge Trading Estate, Weybridge, Surrey. Tel: Weybridge 48311/51466.

RECKITT'S (COLOURS) LTD., Morley Street, Hull. Tel: 0482 29875. Telex 527065.

RUSSELL FINEX LTD., Russell House, Adam Street, Strand, London WC2N 6AE. Tel: 01-836 0055. Telex 21666.

SCC COLOURS LTD., Maryland Road, Stratford, London E15. Tel: 01-534 4151. Telex 263664.

"SACHTLEBEN" CHEMIE GMBH, UK Sales Office, 56 High Street, Edenbridge, Kent. Tel: 073-271 3694. Telex 95297.

SANDOZ PRODUCTS LTD., Calverley Lane, Horsforth, Leeds. Tel: 097-34 4646.

SANDVIK CONVEYOR GMBH., 7012 Schmiden/Stuttgart, Federal Republic of Germany.

SCADO, B.V., Zwolle, Holland.

Synthetic Resins Ltd., Edwards Lane, Speke, Liverpool L24 9HR. Tel: 051-486 3922. Telex 627297.

SCHERING AG, Bergkamen, Federal Republic of Germany.

Schering Chemicals Ltd., Industrial Chemicals Division, The Brow, Burgess Hill, West Sussex RH15 9NE. Tel: 044446-6011. Telex 87577.

SHELL CHEMICALS (UK) LTD., Villiers House, 41–47 Strand, London. WC2N 5LA. Tel: 01-839 9070. Telex 21795.

SPRAY FINISHING SYSTEMS LTD., Hampden Park, Eastbourne, Sussex BN22 9AE. Tel: 0323 52155. Telex 877248.

STEIN ATKINSON STORDY LTD., Ounsdale Road, Wombourne, Wolverhampton, Staffordshire, WV5 8BY. Tel: 0902 4171.

TAM SAMES, 21, rue Jean Macé/38000 Grenoble, France. Tel: (76) 90.32.54. Telex 32171F.

Binks-Bullows Ltd., Pelsall Road, Brownhills, Walsall, Staffordshire WS8 7HW. Tel: 05433-2571/4911/4912.

TIOXIDE INTERNATIONAL LTD., 10 Stratton Street, London W1A 4XP. Tel: 01-499 6070. Telex 22892.

USI ENGINEERING

Leesona Plastics Machinery Ltd., Burtonwood, Warrington, Lancashire WA5 4HB. Tel: 09252 5131. Telex 62530.

U.C.B.

Brandhurst Ltd., PO Box 70, Wellington Road, High Wycombe, Buckinghamshire. Tel: 0494-33411.

VEBA CHEMIE AG., D 4660 Gelsenkirchen-Buer, Postfach 45, Federal Republic of Germany. Tel: (0 23 22) 3861.

Industria (Chemical Services) Ltd., Marbaix House, Bessemer Road, Basingstoke, Hampshire RG21 3NT. Tel: Basingstoke 61881. Telex 858418.

VOLSTATIC COATINGS LTD., 56–67, Stirling Road, Acton, London W3. Tel: 01-992 6931. Telex 934360.

WERNER & PFLEIDERER, D 7000, Stuttgart 30, Theodorstrasse 10, Postfach 30 12 20, Federal Republic of Germany. Tel: (0711) 89561. Telex 722696.

Werner & Pfleiderer (UK) Ltd., 72A Compstall Road, Romiley, Stockport SK6 4DE. Tel: 061-494 1135. Telex 667814.

WINKWORTH MACHINERY LTD., 1 Bridge Street, Staines, Middlesex TW18 4PX. Tel: Staines 55951. Telex 934611.

WOLSTENHOLME BRONZE POWDERS LTD., Springfield Works, Sharples, Bolton, Lancashire. Tel: 0204 52244. Telex 63251.

Appendix 2

SOME USEFUL DATA

Powder Coating Cost per square metre =

$$\frac{\text{Film Thickness (in microns)} \times \text{Specific Gravity} \times \text{Price per kilo.}}{1000}$$

One square metre = 10·764 square feet
Power Usage in grams per square metre = Thickness of Film (in microns) × Specific Gravity.

Sieves

British Standard BS 410-1962	German Standard DIN 4188	U.S.A. Standard ASTM E-11-61	Nominal Size of Sieve Aperture		
			inch	microns	mm
			0·0004	10	
	25			25	
			0·0010	25·4	0·0254
	28			28	
	32			32	
	36			36	
		400	0·0014	37	0·037
	40			40	
		325	0·0017	44	0·044
350	45		0·0018	45*	0·045
	50			50	
300		270	0·0021	53	0·053
	56			56	
240	63	230	0·0025	63*	0·063
	71			71	
		200	0·0029	74	0·074
200				75	
	80			80	
		170	0·0035	88	0·088
170	90		0·0036	90*	0·090
	100		0·0039	100	0·100
150		140	0·0041	105	0·105
120	125	120	0·0049	125*	0·125
		100	0·0059	149	0·149
100			0·0059	150	0·150
		80	0·0070	177	0·177
85	180			180*	0·180
72		70	0·0083	210	0·210
60	250	60	0·0098	250*	0·250
52				300	0·300
44	355		0·0140	355*	0·355
30	500	35	0·0190	500*	0·500
16	1000	18	0·0390	1000*	1·000

* These correspond to sieves proposed as
International Standard ISO/R 565–1967

BS 410–1962	gives the number of meshes per inch of sieve.
DIN 4188	gives the sieve apertures in micrometers.
ASTM E-11-61	gives the number of meshes per inch of sieve.

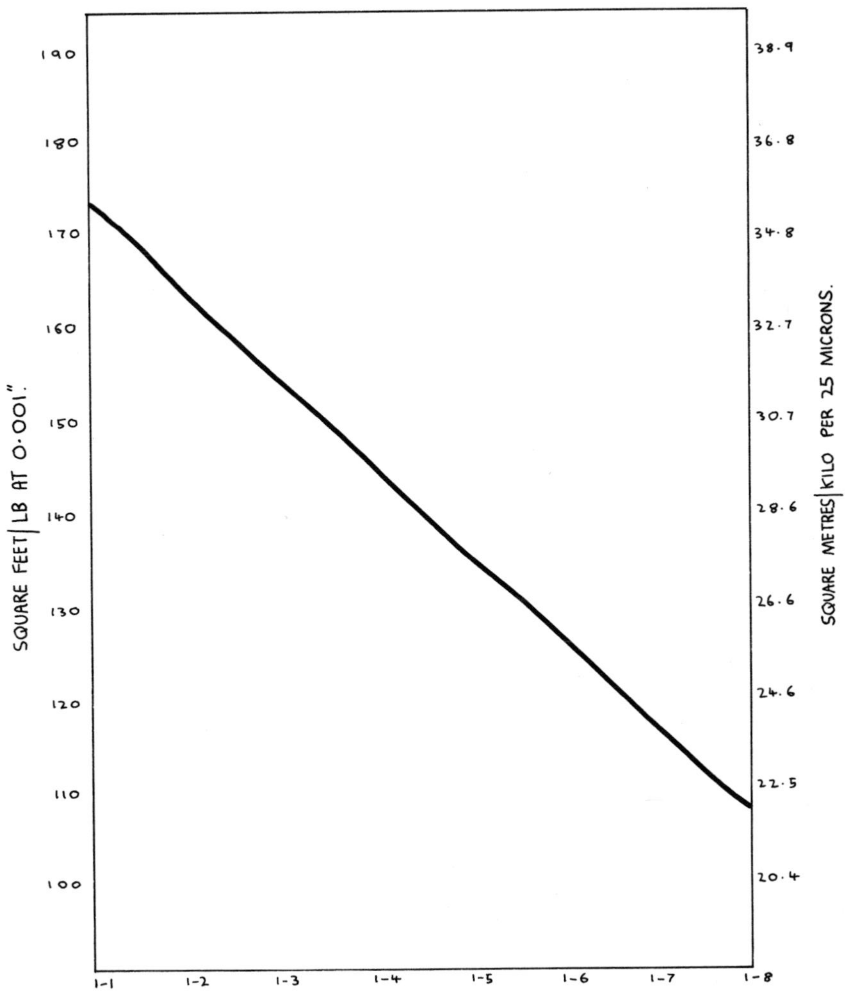

Relationship between Coating Weight per Unit Area and Powder Specific Gravity.

COURTESY DRYNAMELS LTD.

Index